STRATEGIC PSYCHOLOGICAL OPERATIONS AND AMERICAN FOREIGN POLICY

Robert T. Holt and Robert W. van de Velde

STRATEGIC

PSYCHOLOGICAL OPERATIONS AND AMERICAN FOREIGN POLICY

THE UNIVERSITY OF CHICAGO PRESS

Library of Congress Catalog Number: 60-14238

The University of Chicago Press, Chicago 37
The University of Toronto Press, Toronto 5, Canada

Composed and printed by The University of Chicago Press
Chicago, Illinois, U.S.A.

• PREFACE

This book deals with the use of the psychological instrument of statecraft. Had it been written a decade ago, it quite likely would have been entitled *Psychological Warfare.* The authors would not have been altogether happy about that term. They undoubtedly would have suggested several alternatives—international political communication, international information, battle for men's minds, and others—but finally would have accepted with reluctance *Psychological Warfare.*

We have decided to use a new term. There will certainly be those among our readers who will be unhappy with our selection of *Strategic Psychological Operations.* It is reasonable that they should expect some sound arguments, if not sincere apologies, for the introduction of a new term in an area already suffering from an overabundance of half-accepted labels. The reasoning behind our choice combines both positive and negative considerations.

The principal negative consideration was a compulsion to get rid of the words "warfare" and "information." "Warfare" cannot, or at least should not, be applied to the larger concepts involved here. Furthermore, it is "psychologically" unsound—if not definitionally ridiculous—to apply that word to activities conducted in the absence of overt hostilities, or at any time when applied to the activities of a government of a democratic and peaceful people toward the government and people of an allied, friendly, or neutral nation. The term "information" so widely used in our government simply fails completely to cover the situation or to convey any idea of the scope of the operations involved. "Operations," then, is our answer to these negative considerations. We believe "operations" implies a large-scale, continuous process in which are included the functions of planning, coordinating, directing, and acting. No other single word seems to us so aptly to include all these elements on the level and in the magnitude herein considered.

We recognize the negative aspects of the word "psychological," but it has been retained *faute de mieux*. It was once said that the efforts of almost all people concerned with psychological warfare to rid themselves of that distasteful term was reminiscent of the Australian bushman who longed for a new boomerang. For years this chap strove to obtain a new and better implement. Finally his wish was fulfilled, but the result was he drove himself mad trying to throw away his old one. The truth of the matter is that, despite the dislike of many government people for the application of a word connoting little-understood theories to their everyday activities, the word "psychological" is not inappropriate here. As will be seen in the text, the operations *are* psychological in a common-sense usage in that they are designed to provoke certain behavioral responses or attitudes, by bringing about changes in *the meaning that men attach* to objects and events in the material world, rather than by changing directly those objects and events. In keeping this word, too, we hope to maintain the interest of the psychologist and social psychologist in developing further their studies on the nature of the instrument with which we are dealing.

On the positive side is the inclusion of the word "strategic." We use it to imply breadth of scope both organizationally and geographically. It also implies continuation over a long period of time. The use of this word is meant to convey something quite different from "campaign" as President Truman used that word in labeling our overseas information program a "Campaign of Truth." To us the word "campaign" implies a specific course of action designed to last a foreseeable length of time and to culminate in "victory." We see no "victory" to be achieved, but rather a constant striving for "success." The use of "strategic" here also means something quite different from the meaning of that word when it is used in such phrases as "the strategy of terror." This usage seems to us false, for it is used to label what is at most really a tactic. "Strategic" as we use it means something national in its origins, international or world-wide in its effect, and decades long in its implementation.

However, we are trying to do something besides introduce a new term. We are trying to introduce a broader and yet more precise concept of the nature of the psychological instrument of statecraft. In developing this concept we have drawn heavily upon the works of sociologists, psychologists, and experts in mass communication. But these materials are useful only as they are selected and interpreted in relationship to the problems of the development and conduct of foreign policy. The basic problem is one of political analysis.

One of the objectives of this volume has been to bridge the gulf between the various social sciences that have a contribution to make and the gulf between the theory and the practice of psychological operations. In other words, we are trying to relate what has been done by different scholars, each working in the confines of his own discipline and in its specialized terminology, to the practical problems confronting the adviser, the planner, the decision-maker in the conduct of foreign policy.

If we accomplish this purpose, the book should be useful to a number of people. Government officials at the higher policy levels who deal with the problems of United States foreign affairs may find here articulated some of the ideas with which they have necessarily grappled, but often in unclear terms. There are many in the Departments of State and Defense as well as in the United States Information Agency whose tasks may be more clearly defined if our concepts have validity. Whether they agree with us or not, instructors in courses in international relations will, we hope, find both the concepts and the case studies useful in pinning down what has been a neglected area in international politics. Instructors in courses in international communication in schools of journalism may well find this book of value in their courses.

Tables 1 and 3 in chapter iii are reprinted by permission of the publishers from Alex Inkeles and Raymond A. Bauer, *The Soviet Citizen: Daily Life in a Totalitarian State*, Cambridge, Mass.: Harvard University Press, copyright 1959, by the President and Fellows of Harvard College. The chart reproduced as Figure 1 in chapter iii is reprinted by permission of the authors from R. C. Snyder, H. Bruck, and B. M. Sapin, *Decision-making as an Approach to the Study of International Politics*, copyright 1954, by the Organization Behavior Section, Princeton University. Chapter vii is adapted from Robert T.

Holt, *Radio Free Europe*, Minneapolis, Minn.: University of Minnesota Press, copyright 1958, by the University of Minnesota, and reprinted with the permission of the copyright holder.

We are indebted to many people. The ideas of a number of former professors and present colleagues can be found in the pages of this book. Special mention must be made of Professors Harwood Childs, William Ebenstein, Edgar S. Furniss, Jr., Richard C. Snyder, and Harold Sprout. Professors Werner Levi and Roy Carter read the manuscript and made valuable suggestions. Professor Lloyd Short read chapter iv and pointed out some weaknesses which we hope have been corrected.

While we were associated with psychological warfare in the United States Army, many of the main concepts were discussed with Brigadier General William Bullock, Colonel Richard M. Lee, Major Harrison Youngren, James B. Adler, and William A. DePree.

Miss Annie Eggen, Miss Sara Jo Kroll, Mrs. Mary McLaughlin, Mrs. Doris Lake, and Mrs. Evelyn Schumacker have helped in the typing of the manuscript.

The patience and long-suffering and tolerance of our wives were severely tested during the work on this book and were not found wanting. That debt, like the others incurred, can be acknowledged, but never adequately repaid.

• CONTENTS

• PART I

INTRODUCTION

Man's effort to gain his ends through threat or persuasion is as old as history, but a systematic and scientific study of how this can be done in an orderly and predictive fashion is very new. It is really only in the last two decades that scholars in the social sciences have tried to analyze and understand what certain statesmen and military leaders of the past have been able to accomplish by a fortuitous combination of wisdom, instinct, and luck in their use of the psychological instrument of statecraft.

The events of World War II and the circumstances of the Cold War which has followed it gave impetus to a great deal of writing on psychological operations with very little agreement on its major facets. What is the psychological instrument of statecraft? And what is the role of the psychological instrument in the overall pattern of foreign policy implementation? Lack of agreement on the answers to these questions is partially due to the lack of sufficient attention devoted to their answers. This is not to criticize those who have labored in the vineyard before us, nor to belittle the results of their labors. Certainly without their work, ours would have been impossible. But it is our contention that the failure to face up to the above questions has resulted in an unclear, imprecise, and inadequate understanding of the roles and missions of the psychological instrument and of how these can fit into the larger effort of national foreign policy operations.

In Part I we present our concept of the nature of psychological operations. Chapter i is devoted to a general presentation. But the full implications of our concept become apparent only as it is examined in relationship to the problems of policy-planning and to the intelligence function. Policy-planning for psychological operations is discussed in chapter ii and the intelligence function in chapter iii.

If the potential of the psychological instrument of statecraft is to be fully utilized, it is necessary to have the organization responsible for its use developed on the basis of an understanding of the instrument—an understanding of the role it can play in the conduct of statecraft generally and an understanding of the special problems involved in its use. In chapter iv we sketch the major outlines of the kind of organization required in the American structure of government if psychological operations are to be successfully planned and executed.

• CHAPTER I

The Psychological Instrument of Statecraft

As a result of the Second World War the United States found itself on a pinnacle of world power and consequently in a role of world leadership. It was a strange and unfamiliar role, and its acceptance was both reluctant and hesitant. There was little in the history and experience of the nation which provided a basis for understanding and dealing with the crucial decisions that faced its leaders at every turn. Almost immediately it became apparent that the peace for which we had fought was not to be a real peace and that the other Great Power to emerge from the war sought ends quite different from ours. Victory over the Axis did not solve all problems, as many Americans seemed to have believed it would. Indeed, compared with the simple directness of effort necessary to the conduct of actual war, the mounting complexities of the so-called Cold War period presented new problems to the American statesmen almost daily. In retrospect it is remarkable that the United States responded to the challenge with such rapidity and imagination. The Truman Doctrine, the Marshall Plan, and the North Atlantic Treaty Organization were the imaginative foreign policies which marked the abrupt change in more than 150 years of American peacetime isolation.

While new problems arose to plague the government, and as policies were developed to meet the unfolding world situation, students of international affairs were developing new terms and new concepts to describe and help more fully to understand the international circumstances facing the United States. One such concept, developed as part of an effort to provide an intellectual framework for the new policies, was that of "total diplomacy" or "total strategy."[1] The originators of this concept argued that the foreign policy goals of a Great Power could be achieved only if strategies of statecraft were developed which were based on world-wide considerations and included the integrated use of all instruments of statecraft. No major foreign policy goal, they contended, could be achieved efficiently or economically by the use of any one instrument or by the various instruments working independently of one another.

Total strategies have never been fully developed or implemented by the United States. The strategies that have been employed have not provided for the integrated use of the four major instruments of statecraft—the diplomatic, military, economic, and psychological.[2] There are undoubtedly a number of reasons for this. A presidential form of government in which there is a division of power between the executive and the legislature places limitations on the probable amount of integration that can be developed. The "built-in" frictions that exist between Congress and the White House are most evident when the same party does not control both. But even when one party is in firm control, Congress may place budgetary or other restrictions on the implementation of plans developed by the executive. Another reason for the failure to develop total strategies is that the foreign policy decision-making apparatus in the executive branch involves such a multiplicity of agencies with overlapping authority and responsibility. Perhaps less evident, but certainly no less crucial, is the fact that the demo-

[1] E. M. Earle, "Political and Military Strategy for the U.S.," *Proceedings of the Academy of Political Science,* January, 1941, pp. 112–19; F. S. Dunn, "Peace Strategies in an Unstable World," *Yale Review,* Winter, 1948, pp. 226–40; H. and M. Sprout, *Foundations of National Power* (New York, 1951), p. 70; H. D. Lasswell, "Psychological Policy Research and Total Strategy," *Public Opinion Quarterly,* Winter, 1952–53, pp. 491–500.

[2] H. and M. Sprout, pp. 40–42, include a fifth: subversion and sabotage. We would argue that this is not on the same level as the others. Subversion is more likely to be the result of the psychological and/or economic instruments, sabotage to stem from psychological and/or military factors.

cratic process itself inhibits the type of central control which can most readily achieve the coordination necessary to actual totality of action in the implementation of a decision.

But in addition to these organizational and constitutional weaknesses, the United States has failed to develop total strategies for another reason. It has never truly understood the nature of the psychological instrument. Because the instrument has not been understood, it has not been fully exploited. It is with this problem that this book is concerned.

It is not surprising that psychological operations have not been fully appreciated. Traditionally, there have only been three major instruments of statecraft, each with its own distinctive tools, techniques, and modes of operations. The military employs the weapons of physical violence. The economic involves the manipulation of the production and distribution of goods and services in an attempt to achieve certain foreign policy goals. The diplomatic is distinguished by the fact that it is primarily concerned with direct negotiations between official decision-makers from various states. For centuries these were the instruments used by the statesmen of the world. It is not that psychological operations were unknown. Indeed, Linebarger has traced their employment back to antiquity.[3] But two developments which have culminated in the twentieth century have made the psychological instrument one of the major instruments of statecraft. First, the development of mass communications has made it possible to spread information and ideas to great masses of people in countries that have experienced even a small amount of modernization. In the twentieth century for the first time in history, it has become possible for the leaders of one nation to communicate directly with the *peoples* of another nation. Second, and equally important, the twentieth century has seen the expansion of the potential or actual power base of a society. This has been particularly true in democratic states with the continual expansion of the fran-

[3] See Paul M. A. Linebarger, *Psychological Warfare* (Washington, 1954), pp. 1–24. Philip Davidson, *Propaganda and the American Revolution* (Chapel Hill, N.C., 1941), gives a thorough account of the use of the psychological instrument of statecraft in a pre-twentieth-century period. When we talk about the "newness" of psychological operations, we are referring to their conscious, planned, organized, coordinated, and continuing use. The existing broad potential for such use is the result of twentieth-century phenomena.

chise to a larger and larger segment of the population. But even the most totalitarian of the modern dictatorships can be distinguished from its authoritarian predecessor by the fact that the former is apparently forced to solicit the active support of its subjects whereas the latter could exist with mere passive acquiescence. The new-found power of the masses and the ability of a foreign power to speak directly to these masses have now made the psychological instrument of statecraft a major one.[4]

The activities involved in the use of the psychological instrument have been variously named and described. "Political warfare" and "psychological warfare" have been extensively used during and after the Second World War. The British leaned toward the former, and the United States toward the latter. "Propaganda," "The Campaign of Truth," and "war for men's minds," "communications weapon," and many more represent other attempts to define and/or give an emotional stamp to the activity in question.[5] Officially, the U.S. operations in this area have been labeled "information" and that word has appeared in almost every one of the several titles which our national propaganda agency and its ancestors have borne. This superfluity of names indicates general lack of agreement on the exact nature and role of the psychological instrument. One American scholar using the term "psychological warfare" to apply to the use of the psychological instrument generally, paraphrases Mark Twain in commenting, ". . . psychological warfare is neither psychology nor war,"[6] while

[4] Another point has been made by Lasswell. He argues that one development since the end of the Second World War has further increased the importance of the psychological instrument. The bipolarization of the world, he says, has limited the ". . . grand strategy of encirclement. Instead of the incessant realigning of national blocs in maneuvers of encirclement and counter encirclement, the principle strategy becomes that of penetration, which puts a premium on dividing an elite against itself, and from the rank and file." In this situation psychological operations become even more important while the value of diplomacy is somewhat depreciated. See Lasswell, p. 494.

[5] Roland Perusse has listed eighteen terms that are sometimes used synonymously with "psychological warfare." See "Psychological Warfare Reappraised," in William E. Daugherty and Morris Janowitz (eds.), *A Psychological Warfare Casebook* (Baltimore, 1958), pp. 25–34—hereafter referred to as *Casebook*.

[6] Saul K. Padover, "Psychological Warfare in an Age of World Revolution," *Columbia Journal of International Affairs*, V, No. 2 (Spring, 1950), 3.

another, equally astute and experienced, has a more restricted point of view, defining psychological warfare essentially as the "application of parts of the science called psychology to the conduct of war."[7] Another feels that the distinctive characteristic of psychological operations is that they attempt to influence the *will* to conduct policy as distinguished from the *capability* to conduct policy.[8] A fourth emphasizes the more strictly propaganda function of "talking" about the activities carried on by the other instruments of statecraft.[9]

The failure to develop an adequate and widely accepted concept or name for the operation of the psychological instrument has been matched by a tendency in official Washington to shun direct responsibility for its use. Note the successive changes in status as well as in name of this governmental function. Note the succession of heads of the agency charged with the function, and of presidential advisors in the area. The details of this change and turnover will be dealt with in a later chapter. Suffice it to say here that there still exists in Washington a peculiar contradiction between an ardent lip service to the objective and a shunning of the tasks necessary to the objective's attainment. Part of this contradiction stems, perhaps, from the terms employed to describe the task. "Propaganda" awakens automatic memories of Herr Doktor Goebbels, and "warfare," whether "political," "psychological," or "communications," is inappropriate and distasteful to those who may be employing this process in an attempt to achieve a real peace.

But not only have some of the words employed been distasteful, they also indicate that the nature and scope of the activities involved have not fully been appreciated. The limits imposed by the use of any term employing the word "warfare" have already been pointed out. But the innocuous word "information," which has gained respectability, if nothing else, also indicates severe limitations in the concept of the nature of the psychological instrument of statecraft.[10] This comes out

7 Linebarger, p. 25.

8 Hans Speier, "Psychological Warfare Reconsidered," in D. Lerner (ed.), *Propaganda in War and Crisis* (New York, 1951), p. 465.

9 D. Lerner, *Propaganda in War and Crisis*, p. xiii.

10 In 1959 a senior official in USIA in a letter to one of the authors disclaimed for his agency the role "wielder of the psychological instrument of statecraft." And under the present charter, he is quite right.

clearly in an examination of the official purposes of the various information agencies that have been established.

The language of the basic law for the U.S. "foreign information program," Public Law 402 (80 Cong.), *The United States Information and Educational Exchange Act of 1948*,[11] states: "The Congress hereby declares that the objectives of this Act are to enable the Government of the United States to promote a better understanding of the United States in other countries and to increase mutual understanding between the people of the United States and the people of other countries." Five years later, the President in his request[12] for the establishment of a separate United States Information Agency, said that the new agency is to be "for the conduct of our information programs." These, he said, "include, with certain limited exceptions, four programs; the information activities now administered by the International Information Administration of the Department of State; the information programs financed in connection with the government in occupied areas; the information program of the Mutual Security Agency; and the Technical Cooperation Administration information program." The monotonous repetition of the word "information" in agency titles and in describing their functions is representative of the narrow interpretation and lack of understanding of the psychological instrument of statecraft which has made Washington incapable of properly exploiting this twentieth-century tool. A further indication of this lack of grasp of the inherent potentialities of this instrument is the frequency with which "official" Washington as well as the general public for some time confused the entire program with, and gave to all of it the title of, *The Voice of America.*[13] The President and the National

11 Often referred to as the Smith-Mundt Act.

12 H.R. Document No. 156, 83 Cong., 1 sess., "Message from the President of the United States Relative to the Conduct of Foreign Affairs," referred to the Committee on Government Operations on June 1, 1953, pp. 3–4.

13 Even a State Department witness, claiming eight years service and pleading with a Senate committee not to separate the "information" function from State, repeatedly called it "The Voice of America," when from the context it is quite apparent that he meant the entire program. See H. R. Report 844, Part 2, 83 Cong., 1 sess., "Additional Views of George Meader" (July 16, 1953), establishing U.S. Information Agency (Reorganization Plan No. 8 of 1953), pp. 3–4.

Security Council prescribed, late in 1953, that the mission of the newly created United States Information Agency would be ". . . to submit evidence to the peoples of other nations by means of communication techniques that the objectives and policies of the United States are in harmony with and will advance their legitimate aspirations for freedom, progress and peace."

On July 8, 1953, the White House released a statement on a highly classified report of the President's Committee on International Activities.[14] Judging from some of the wording of this statement, the committee seemed to come close to understanding and recommending the use of the psychological instrument. It said, "In reality, there is a psychological aspect or implication to every diplomatic, economic or military policy or action. This implication should receive more careful attention, both in the planning and execution stage of policy, but not to the exclusion of other factors." But then the statement goes on to say that "except for propaganda there are no 'psychological warfare' instruments distinct from traditional instruments of policy." One is tempted to wonder if the committee did not find that the only ready "solution" to an admittedly complex and trying problem was to deny the existence of facts which gave us the problem in the first place. It did bow to the realities of international relations to the extent of saying that the important thing was "to coordinate and time such [day-to-day governmental] actions so as to derive from them the maximum advantages."

Later in the report the "information" process was treated at some length, but was almost entirely divorced from policymaking. Only a reportorial role was left to the "information" agency. The nation once again missed an opportunity to realize its full potential in the arena of international relations because the government deliberately denied itself the full use of the psychological instrument.

The mandate of the United States Information Agency enables it to report on and explain our foreign policy and its implementation. As a corollary, of course, it should "explain" the

[14] Statement published in the *New York Times*, July 9, 1953. The report in question later came to be known in government circles as "The Jackson Report," named for the committee's chairman, William H. Jackson.

United States and "sell America." In other words, its mission is one easily understood by most twentieth-century Americans —that of public relations man and advertising agent for the nation. But at best this is only part of the job. In some parts of the world where the citizens are immune to the blandishments of "institutional advertising" and too closely concerned with vital problems of daily existence to listen to language better suited to say that the United States is "kind to the hands," or glistens because its people use a lanolin hair lotion, something quite different is needed.

It is a basic thesis of this book that there *is* a "psychological warfare" instrument distinct from the three more traditional instruments; that this instrument *is* as analytically separable, and as operationally inseparable, from those instruments as, say, the diplomatic is from the military. Reporting and explaining government actions is one function of that instrument; helping to coordinate and time "such actions so as to derive from them the maximum advantages" is another of its functions. But to be fully exploited the psychological instrument must also have a part in initiating policy and action. This is not to claim any position of pre-eminence for the psychological instrument; it is to plead for it a status of equality with the diplomatic, economic, and military instruments, each of which will find its role constantly moving up or down a scale of relative importance, reflecting the set of circumstances peculiar to any given world situation in response to which the particular policy must be developed or the particular action taken.

It is on this ground that the present writers find themselves in sharp disagreement with many other writers in this field who appear content with both the limited role thus far ascribed by the United States to "psychological warfare" or "information program," and content also with these self-limiting terms for that role.[15]

[15] There are numerous examples in the Daugherty and Janowitz *Casebook*. And although Daugherty does say, in the article, "Changing Concepts," pp. 12–18, "There is certainly great need for a new term, but before an acceptable one can be coined it is first necessary that there be greater agreement as to scope of the activity that has been described as 'psychological warfare,'" the tone of the editors' introductory chapter is one, if not of acceptance, at least of acquiesence in the role and the term. This really monumental work in the

In order to introduce our concept of psychological operations it is necessary to discuss briefly here our view of the nature of statecraft. All foreign policies have an objective (or objectives) and a strategy. The objective is an image of a future state of affairs viewed in relationship to some present state of affairs. The strategy is a plan of action which spells out the manner in which the instruments of statecraft with their various tools and techniques are to be used to achieve the objective. The strategy of foreign policy is always aimed at achieving certain desired responses from the state or states being acted upon.[16] For instance, it was apparent that the major objective of the Truman Doctrine was the creation of a state of affairs in Greece and Turkey in which the influence of the Soviet Union was minimal compared to the influence exercised there by the United States. The strategy (which called primarily for the use of military, economic, and technical aid together with some show of the American "flag") was designed to get the Soviet Union, Greece, and Turkey to act in certain ways believed by the United States to be in its national interest.

If one desires to develop a plan of action to influence the behavior of foreign states, he must have an idea of what determines that behavior. Snyder, Bruck, and Sapin point out that the "key to the explanation of why the state behaves as it does lies in the way its decision-makers define their situation."[17]

To influence the behavior of a state it is necessary to develop a strategy aimed at affecting the manner in which its decision-makers define the relevant situations.[18] Traditionally, strate-

---

field is, of course, a casebook, and as such may not be appropriate as a vehicle for the editors' ideas. On the other hand, the opening remarks and the title they chose would seem to put them in at least the most recent lead of those with whom we disagree in basic concepts.

16 This analysis of policy is developed in greater detail in chap. ii.

17 R. C. Snyder, H. W. Bruck, and B. Sapin, *Decision-making as an Approach to the Study of International Politics* (Princeton, 1954), p. 51—hereafter referred to as Snyder *et al., Decision-making.*

18 We recognize that the strategy to influence the behavior of a foreign state may involve initially an attempt to overthrow the existing decision-makers. The assumption would be, of course, that the new decision-makers would pursue a policy that would be more acceptable than that of the old leaders.

gies employing the economic, military, and diplomatic instruments of statecraft have attempted to influence foreign decision-makers' definition of the situation by materially changing the situation. Defense expenditures may be increased, an army may be moved up to a troubled border, an embargo may be placed on the shipment of certain goods. Diplomatic negotiations may lead to the formation of a new alliance. As the situation is changed, decision-makers' definition of the situation will likely change (although there almost certainly will be no one-to-one relationship between changes in the situation and the way in which various decision-makers define these changes).

But it is not only the decision-makers' definition of the foreign situation that affects their formulation of policy. Domestic considerations may also affect foreign policy decisions. Therefore, in analyzing foreign policy decision-making, it is necessary to distinguish between the decision-makers' definition of the domestic and foreign situations and to understand the relationship between them. If one's object is to precipitate certain kinds of behavior, this can be done by affecting the decision-makers' definition of either their internal or their external situation. Historically, emphasis has been placed on the external situation for the simple reason that it has been more difficult to manipulate objects and events within another state. There are, of course, some notable exceptions. The peacemakers at Vienna in 1815 were greatly concerned with the problem of legitimacy and saw as one aspect of their responsibility for keeping the peace the support and maintenance of "legitimate" governments in Europe. Germany very successfully influenced the foreign policy of Russia in 1917 by manipulating the internal situation when it sent Lenin into that faltering country.

But these examples to the contrary notwithstanding, the traditional methods for a state to influence the behavior of a foreign state have been through the manipulation of material objects, usually in the external environment of the state being acted upon. These traditional strategies involved military and/or economic operations and diplomatic negotiations.

The distinctive nature of psychological operations derives from the fact that they do not attempt to manipulate material objects, but rather try to influence the way in which material objects are perceived and interpreted.

To make this distinction more precise, one can differentiate between the world of objects and events and the way in which objects and events appear to individuals and have meaning—in other words, to distinguish between the world as it "is" and the world as it "appears" and has meaning.[19] The former we will call the *material* world and the latter the *apparent* world. They are not necessarily the same. The dramatic experiments in perception carried on by Ames and Cantril have shown that even such familiar objects as a window or a chair can be greatly distorted if they are seen in unusual surroundings, such as in a room with a sloping floor and trapezoidal walls. In such a room men appear to be either giants or pygmies, depending on where they stand in relation to the viewer's position.[20]

The differences between the material world and the apparent world are increased and complicated if action and time are introduced into a situation. One has only to note the different accounts eyewitnesses give, quite honestly, to an accident at which they were present. Obviously a set of "facts" exists. Two cars of certain make and model, say, approached each other at certain speeds at a certain time on a certain day and collided in a certain manner at a certain spot with certain resultant damage. This is the material world—the world as it is, or was. But both drivers, even assuming complete honesty on the part of each, may have very different versions of what happened. Furthermore, the drivers' accounts may be quite different from those of two witnesses who happened to be standing at the side of the road at the time. What is more, even the accounts of these identically placed, disinterested witnesses may vary markedly in at least some important respects. This simple example shows two things. First, that the material world and al-

19 Obviously when speaking of the world as it "is" we are not referring to a world which exists apart from any human observation, for that would be a purely metaphysical concept. The distinction we are making is similar to that made by F. H. Allport. (See *Theories of Perception and the Concept of Structure* [New York, 1955], pp. 20–46.) Allport differentiates between phenomenological experience (the world as it appears) and physicalistic experience (or the world as it "is" as determined by denotation or encounter).

20 The role of the so-called "new" theories of perception on our concept of the use of the psychological instrument will be discussed in chap. iii.

most anyone's apparent world are often not the same and, second, that even in small, familiar situations actually witnessed, apparent worlds may be quite different from person to person.

The differences between the physical world and the apparent world become even greater, of course, if the situation being "viewed" is more abstract or complex. For example, it is highly unlikely that any two members of the President's Cabinet look on the "Middle East situation" in exactly the same terms. Yet it is likely that their apparent worlds in that area are closer to each other than any of them would be to the situation there as seen by a New York City steelworker or a Las Vegas bartender. Again, it is likely that all these Americans' views of the situation would bear more resemblance to each other than any of them would to the views held, for example, by the Deputy Minister of Defense in the Kremlin or by the Prime Minister of Egypt.

This oversimplified discussion is to emphasize that not only do the material and apparent worlds differ, but that each person's apparent world is, in large measure, delineated by his past and present cultural, educational, geographic, and occupational milieu. These two points—that the material and apparent worlds are not the same, and that individual or group apparent worlds may vary greatly—are fundamental to an understanding of the nature and importance of the psychological instrument of statecraft.

Equally important is a realization that while changes in the material world usually cause changes in the apparent world, there may be no one-to-one relationship. It is even quite possible to have a change in the apparent world without any change in the material world. Changes in either can affect behavioral responses. If a man's house is destroyed by an enemy bomb, the material world and most likely his apparent world will be changed and these changes will evoke some response in behavior and attitude. Part of the response will be "caused" by the material change; he will have to find a new place to live. But the meaning he ascribes to that change—or his new apparent world—will also affect his behavior. He may shake his fist in defiance and work doubly hard to help destroy the enemy. Or he may say to himself, "What's the use? We might as well quit before everything is lost." Changes in the apparent world can come about by various means without changes in the material world. The process of education does it daily to children, and it has been said that no one is too old to learn. The stereotypes one has of foreigners and foreign places—a manifestation of one's apparent world—are often very "real" until there is an opportunity to travel. This is not to say that firsthand observation will cause one's apparent world to coincide with the material world. It almost certainly will not. But travel will often cause a change in the apparent world without any change in the material world at all. Again it must be stressed that the changes caused are not the same, or equal, or even in the same direction, in two persons making the same trip at the same time. What one "sees" is conditioned by what environment he grew up in, how far he went in school, what his interests are, what he *expects* to see, and what he *wants* to see. It cannot be stated too emphatically that it is man's apparent world, *not* the material world, which determines his attitude and his behavioral response to any given situation.

This discussion provides the basis for distinguishing the various instruments of statecraft and establishing the unique characteristics of the psychological. The diplomatic instrument is identified by the fact that it involves direct contact between the officials of one state with the officials of another. Many times these contacts are negotiations that lead to changes in the material world. The economic and military instruments of statecraft involve direct manipulation of the material world. The psychological instrument involves manipulation of the apparent world.

The question, "Whose apparent world?" will be discussed in the next chapter. We shall just point out here that ultimately it must be the apparent world of the foreign decision-makers or what we have called their definition of the situation. But this may be achieved indirectly. Propaganda broadcasts to a foreign population might cause many people to change their attitudes toward their government's policy, which in turn might be instrumental in the decision-makers' estimation of their domestic support, which in turn could lead to a policy change. Psychological operations in a third country might contribute to a change in leadership, which in turn requires the decision-makers in other states to redefine the external situation. But whenever the initial operation involves an attempt to manipulate peoples' views of the world rather than material objects and events, one is dealing with the psychological instrument of statecraft. It should become increasingly apparent in the next two chapters that this concept of psychological operations involves something more than information, or "talking" about policies, or "influencing the will to conduct policy."

One further point should be made. Obviously diplomatic, economic, and military operations do have psychological implications and in many cases it is most difficult to label an operation as *purely* economic or *purely* psychological. It is more useful to think of operations that are woven together in a strategy to affect the behavior of foreign states as *primarily* eco nomic, *primarily* psychological, etc.

• CHAPTER II

Policy and Planning

Foreign policy-making in a modern-day Great Power is a most complex process. The agencies involved, if one includes the geographic and functional subdivisions of the major departments, number in the hundreds, the personnel in the hundreds of thousands, and the financial costs in the thousands of millions. A man on the inside or an acute observer on the the outside may have a reasonably complete picture of the daily operation of a small fraction of this widely ramified process. In addition such a person might also carry in his mind's eye a picture, in the most general outlines, of the process in its entirety. But no one does—indeed, no one can—know in detail the whole operation. Even if we limit ourselves to policy-making for psychological operations, a comprehension of the complete process with all its richness of detail is outside the compass of a small volume—or of a series of large volumes, for that matter. But if one could distinguish the significant from the unimportant aspects of the policy-making process, this impossibility of complete knowledge would not be of great consequence.

It is one of the functions of a frame of reference to do precisely that—to provide a way of thinking about a subject which identifies the significant variables and builds a foundation for interpreting how they interact. In reality, anyone viewing the policy-making process with an eye to understanding it has a frame of reference. It may be implicit; indeed, the observer might not even be aware of it, but he does have some way of ordering and interpreting the various information he receives. It is useful for the policy-maker and for the student to have an explicit frame of reference so that its validity can be checked and it can be modified to perform its function more adequately. We are going to present our discussion of policy-making for psychological operations in an explicit frame of reference which we hope will provide some useful checkpoints both for the student and for the policy-maker.

A Frame of Reference for Viewing the Policy-making Process

Let us start from the very beginning and define what we mean by a policy. A *policy* is a statement of an objective (or objectives) together with a broad plan of action designed to reach the objective. In our frame of reference it is useful to distinguish between the two aspects—the *objective* and the plan of action which will be called a *strategy*.

OBJECTIVES

It is important also to be explicit about the definition of an objective. There are a number of synonyms for the term—goal, purpose, mission, end, etc. But synonyms tell one very little. We follow that group of writers who define an objective as "an *image* of a future *state of affairs*—a *set of conditions* to be fulfilled or a *set of specifications* which when met are to be regarded as achievement of what was desired by the decision-makers."[1] This definition has some important implications. It is not very meaningful to talk about any *future state of affairs* without reference to an *existing state of affairs* which the policy-makers desire to modify or preserve. We shall call this existing state of affairs a *situation*. Situations and, therefore, objectives can be identified in terms of geography, time, or issue. Thus one may speak of American objectives in Europe as contrasted with those in the Far East, or of objectives for the next six months as contrasted with those of the next decade, or of objectives on the disarmament issue. It is also possible to conceive of objectives that combine two or more of these three in one manner or another.

The three factors of geography, time, and issue each contain

[1] Snyder *et al., Decision-making,* p. 51.

many variables, and each of these may be multiplied by having to be considered in the light of the variables of either or both of the other factors. For instance, we may have to determine U.S. objectives in France as against those in Algeria on the basis of today and ten years from now, measuring these against our hope of achieving an East-West arms limitation agreement while not jeopardizing the strength of NATO. It is obvious from this quick and limited example that a state "exists" in many "situations" at one and the same time and therefore must have many objectives. These objectives are not all of equal importance, nor are almost any other objectives one could conceive of, but they, like all others, *are* interrelated. This interrelation of objectives makes it useful to look at them as forming a hierarchy or pyramid.

The objective at the very top of the pyramid may be labeled the *national interest.* This is an ambiguous term and it is perhaps impossible to define it rigorously and identify its empirical content in any way that would have widespread agreement.[2] This is not to say that it is an unimportant goal. The problem is that various policy-makers often have different ideas of just what the national interest is. And not only may different decision-makers have different concepts of the national interest, they also rarely state clearly in concrete terms exactly how they conceive the national interest. That is, they may speak of the national interest in terms of *freedom* or *world peace* or *security*, but they rarely spell out clearly what kinds of images of the future are involved. At the lower levels of the pyramid, however, objectives are usually stated in more concrete terms and there is more agreement on what the objective is if not necessarily on its desirability. Official statements made it relatively clear, for example, that the objective of the Truman Doctrine was to prevent the spread of Soviet influence in Greece and Turkey or, in our terms, to create an image of the future in which the Dardanelles remained within the Western sphere of influence.

Two further points should be made about this conception. First, in this pyramid the objectives at the lower levels are also the means to achieve those goals which are at the upper

[2] Cf. Arnold Wolfers, "National Security as an Ambiguous Symbol," *Political Science Quarterly*, December, 1952, pp. 481–502.

levels. One of the "lower level" goals of the Marshall Plan was to raise living standards in Italy and France. It was felt that this objective would in turn lead to a reduction of the appeals of internal communism, which in turn was a means of helping to maintain Western-oriented decision-makers in power, which in turn was believed to serve the U.S. national interest. The objectives viewed *from below* appear to be goals to be accomplished, but viewed from above, these "objectives" become "means," and this is true at any given level.

Second, looking laterally across the pyramid at any given level, the objectives may appear more or less incompatible—particularly from the point of view of the people responsible for the achievement of a given goal. The prosecution of the war in Korea was partially incompatible with the maintenance of a firm security position in Europe. This kind of incompatibility is to some degree inevitable. Resources are limited and those used for the attainment of one objective cannot be used for the attainment of another. There is another way in which objectives may be incompatible. The strategies necessary for the attainment of one objective may preclude the attainment of another. With the great diversity of interest in the world, it is difficult to conceive a situation in which the strategies to achieve the foreign policy goals of a Great Power vis-à-vis one state or geographical area would not affect the achievement of its goals in another area. For example, the postwar reconstruction of German industry certainly was incompatible in some respects with helping to rebuild British markets overseas.

The necessary fact of incompatibility leads to a discussion of another characteristic of objectives. All objectives must be given a priority rating. But this priority rating is meaningful only in relationship to the time dimension. Some objectives which have high priority over the long run must be subordinated for short periods of time to a less crucial objective which momentarily may be more important. For example, the situation in South Korea in June of 1950 and American objectives there following the invasion were not generally considered to be as important in the long run as the situation in Western Europe. However, troops and material originally destined for Europe were rerouted to Korea, because failure to take *immediate* action in Korea might have precluded the possibility of any effective actions. So for a short period of time the U.S. security position in Europe was weakened somewhat in order to act at a critical moment in a situation of lower long-run priority.

THE SELECTION OF OBJECTIVES

In any given situation there are a number of possible objectives from which policy-makers must choose one. How is this choice made? It is possible to postulate that two criteria are of predominant importance in the selection process—the criterion of *desirability* and the criterion of *feasibility*. The policy-making process is partially the process of choosing the optimum objective from among those alternatives that have been assessed according to how desirable and how feasible they are. One can conceive of the policy-making process first of all as involving implicitly or explicitly a rank ordering of the various alternatives on the basis of these criteria. The ordering on the basis of desirability is done in relationship to the policy-maker's view of the national interest. We must point out again that various policy-makers may have different views of the national interest. Some may emphasize the major values of the society as conditioned by historical and cultural experience. Others may tend to define the national interest more in terms of some amalgam of various special interests. Thus different policy-makers may have different desirability rankings. But the unit that is responsible for making a certain decision will have to come to an operating agreement or stagnation ensues.

It might be said figuratively that the ranking of the various alternatives according to the criterion of desirability involves looking "up" the pyramid of objectives. The possible alternatives are ranked in order of the degree to which they contribute to the achievement of the national interest.

The criterion of feasibility, as the term suggests, is related to the degree to which the various alternatives are considered possible. But estimates of feasibility can be made only in terms of an examination of the action plans or strategies that may be employed to achieve the alternative objectives. Thus a choice of objectives cannot rationally be made until at least the general outlines of the tactics and strategies to achieve them are worked out and examined. There are three kinds of information or knowledge that are necessary in the development of a tactic or strategy. First, there must be an analysis of the situation; second, there must be a doctrine concerning the uses of the various instruments of statecraft in the situation; third, there must be an estimate of the domestic limitations imposed on the means available for use in the situation.

This frame of reference has been presented in a most summary fashion, but it provides a basis for presenting some ideas about policy-making for psychological operations. There is no inherent reason why policy-makers who are primarily concerned with the use of the psychological instrument of statecraft should have by virtue of that fact any unique concept of the national interest. Furthermore, the determination of the desirability of various objectives is likely to occur at such a high level of government (in the United States at the National Security Council level or higher), that this function can hardly be looked upon as a proper topic for this book. Our concern is with the development of strategy and the feasibility aspect of the selection of objectives.

One of the three kinds of information or knowledge required for the development of strategy, the analysis of the situation, will be treated in the next chapter, on the intelligence function. This chapter will take up the problem of doctrine and the matter of domestic limitation.

A Doctrine for the Conduct of Psychological Operations

Policy-makers must have a doctrine concerning the use of any instrument of statecraft. It is clear from an examination of the negotiations between representatives from the United States and the Soviet Union that quite different views prevail in these two governments on how to use the diplomatic instrument of statecraft. Also some of the differences between the Western powers on how to negotiate with the Soviet Union result from different doctrines of diplomacy. However, it is in regard to the wartime use of the military instrument of statecraft that the most explicit doctrines have developed. Military establishments have been extremely articulate and detailed about these matters. Each service, branch, and corps has its mission and there is a body of tactical and strategic doctrine concerning mass, maneuver, and shock action. A history of military operations reveals the importance of an adequate doctrine. Many times forces employed on the basis of a superior doctrine have defeated forces with superior material capability. The German victories on the western front in 1940 against superior numbers and at least equal equipment are only the most recent of a long series of historical examples.

There are a number of bodies of doctrine concerning the use of the psychological instrument of statecraft. The Nazis had perhaps the most explicit doctrine,[3] but there is also a body of doctrine underlying the operations of the USIA, although much of it is implicit and unclear.[4] R. H. S. Crossman, who has been generally acclaimed as the most successful propagandist on the Western side during the Second World War has spelled out in some detail his ideas on doctrine concerning the use of propaganda against a totalitarian opponent.[5] But none of these are based on a very useful notion of what the psychological instrument of statecraft is and what the nature and function of a doctrine is.

In chapter i we argued that all foreign policy strategy is a plan of action designed to influence the behavior of foreign states. A large number of alternative approaches may be used, including the attempt to bring about a change in the foreign decision-makers either peacefully or through a revolution. But all the possible alternatives must be directed at getting foreign decision-makers to define the relevant situations in a manner that will lead to their taking certain kinds of action. It is the function of a doctrine for the conduct of foreign affairs to spell out the ways in which the tactics and techniques of the instruments should be employed in order to precipitate certain behavior from a foreign state under a given set of circumstances. In order to develop an adequate doctrine concerning the uses of the psychological instrument of statecraft, it is necessary to have a clear idea of what the instrument is and of what techniques and tactics are available. We argue that the distinctive nature of the psychological instrument of statecraft derives

3 Daugherty and Janowitz, *Casebook*, pp. 47–54.

4 L. Bogart, "Operating Assumptions of USIA," *Public Opinion Quarterly*, Winter, 1955–56, pp. 369–79.

5 Daugherty and Janowitz, *Casebook*, pp. 35–55.

from the fact that it attempts to influence the behavior of foreign states, not by manipulating objects in the material world, but by affecting the manner in which the world is perceived and interpreted, whether or not there has been any actual material change in the situation.

Analytically it is possible to establish four ways in which psychological operations can influence the behavior of foreign states. Two have to do with the situation external to the foreign decision-maker, and two deal with his internal, or domestic, situation. Two are designed to affect him directly, and two are intended to achieve their purpose indirectly. A listing would appear as follows:

A. External (or "foreign" to the foreign decision-maker)
   1. Affect directly the manner in which he perceives the external situation independently (or relatively independently) of any actual changes that take place.
   2. Actually modify the external situation of the foreign decision-maker by manipulating the apparent world of people outside his national boundary.

B. Internal (or "domestic" to the foreign decision-maker)
   1. Affect directly his perception of some aspect of his own nation which has a bearing on his choice of foreign policy alternatives.
   2. Modify the attitudes or behavior of the populace, or that portion of it, on which the decision-maker's power depends.

It may be helpful to a clearer understanding of these approaches if an example is attempted for each. So far as one not privy to the whole story can tell, it would appear that the "apparent world" view of their nations' international situation, held by the French and British leaders at the time of the Suez venture in 1956, was changed in mid-course by a Soviet threat and/or an American protest directed specifically at those leaders for the very purpose of causing them to take the action they took in halting the military operation. The Soviet "atomic blackmail" threat was unquestionably a psychological tactic, whether or not they would have backed it by action had their public threat not been heeded. The American protest, which was directed to the European leaders in normal diplomatic channels, might have remained purely a manifestation of the diplomatic instrument had it not been also publicly "released." When this was done, the tactic became "psychological" in that the publicity itself was obviously intended to influence the British and French leaders' thinking. (The fact that the "message" was also intended for Russian minds does not change the intent in regard to the British and French. Indeed, it was probably related to that intention.)

As an example of indirectly affecting the external view of a decision-maker, one might cite any number of possibilities. For instance, if Soviet propaganda and agitation in Iran should help bring about the establishment of a pro-Soviet government in that country, there would be a change in the material world, which would undoubtedly affect U.S. policy-makers' view of the situation in the Near East. This would most likely precipitate a change in American foreign policy. The initial phase, however, would involve the Soviet use of psychological operations in Iran.

It has been suggested by Speier[6] that it would not have been difficult for British psychological operators to change Hitler's view of the internal German situation if, after Rudolf Hess's dramatic 1941 flight to England, they had suggested that there were other high German officials contemplating similar escapes. It is conceivable, anyway, that this might have set off a purge in top-level Nazi circles which would have had serious implications for the German war effort.

The fourth way in which psychological operations can effect policy change is by affecting those on whom the decision-maker's authority depends. For example, there is no doubt that Egyptian propaganda and agitation helped bring about the revolt within Iraq in 1958, with consequent changes in official Iraq's attitude toward Egypt as well as toward East and West. Psychological operations can help build up or destroy morale in peace as well as in war. They can help create support for a particular group in a society or for a particular course of action; they can equally well help to dissipate such support as already exists. These changes in the internal situations should help bring about changes in the behavior of the policy-makers.

[6] Hans Speier, "Psychological Warfare Reconsidered," in D. Lerner (ed.), *Propaganda in War and Crisis* (New York, 1951), p. 485.

All four of the approaches to the use of this instrument of statecraft involve, in the first instance, the manipulation of meaning rather than objects or events—manipulation of the apparent world rather than the material world. Furthermore, they include a range of activity which is far greater than "selling America," or "making friends for democracy," or "exposing communism," or "battling for men's minds." It should also be emphasized, however, that this increased range of activity is of little consequence unless it is also recognized that, no matter who may be the initial focus of operations, the purpose is to influence the behavior of foreign states in certain desired ways.

In order to develop a doctrine one must have not only a concept of the nature of the instrument but also an understanding of the techniques and tactics that may be employed. There are four techniques and four tactics that are available. The techniques are propaganda, rumor, direct person-to-person communication, and symbolic acts. The tactics are deception, enlightenment, terror, and reassurance.

TECHNIQUES

Propaganda.—There are almost as many definitions of propaganda as there are writers on the subject. Doob calls propaganda "the attempt to affect the personalities and control the behavior of individuals toward ends considered unscientific or of doubtful value in a society at a particular time." Bartlett defines propaganda as an "attempt to influence opinions and conduct—especially social opinions and conduct—in such a manner that persons who adopt the opinions and behavior indicated do so without themselves making any definite search for the reasons." In the well-known Smith, Lasswell, and Casey volume on communication, propaganda is considered to be "language aimed at large masses" for the purpose of influencing "mass attitudes on controversial issues."[7]

Much of the literature dealing with the definitions of propa-

[7] B. L. Smith, H. D. Lasswell, and R. D. Casey, *Propaganda, Communication, and Public Opinion* (Princeton, 1946), p. 1. These writers specifically exclude from propaganda words used "in diplomacy" and words "used in teaching how to read, write and figure." See also L. W. Doob, *Public Opinion and Propaganda* (New York, 1948), p. 348; F. C. Bartlett, *Political Propaganda* (Cambridge, 1940), pp. 5–6.

ganda is concerned with meaningless issues—hairline distinctions between "education" and "propaganda" or discussions of whether or not propaganda is "true" or "false," "good," "bad," or "neutral." We shall define propaganda for our purposes simply as the attempt to influence behavior in the direction of some relatively specific and explicit goal by affecting, through the use of mass media of communication, the manner in which a mass audience perceives and ascribes meaning to the material world. Three aspects of this definition deserve some elaboration. First, a mass audience is involved. A man trying to convince his friend that he should vote a certain way in a coming election is not using propaganda. The dynamics of this highly personal contact are different from the dynamics involved in the use of the mass media and the instruments and organization necessary to use the mass media are different from those used even in an ambitious program of personal agitation.[8] But mass communications do not have to be directed at a large audience or at an amorphous audience. By the special use of language, format, time and place of dissemination, etc., it is possible to select specific groups as major audiences for mass communication. But the audience is mass in the sense that there can be no discrimination between the individuals who may make it up. *Harper's Magazine* is designed for a middle- and upper-class audience with more than an average education. It is likely that a majority of its readers come from this relatively small group. However, it is impossible to insure that the journal will be read by John Doaks and not by Jim Doe. In a similar vein Captain Zacharias' broadcasts[9] to the Japanese in 1945 were directed at Japanese leaders or potential leaders—a small and select group, some of whom he knew personally. Yet his broadcasts may be called propaganda because he could not address leader A to the exclusion of leader B, nor could he prevent non-leaders from hearing the messages.

A second point that must be emphasized is that propaganda

[8] This is not to say, however, that propaganda messages cannot be carried through a word-of-mouth network. Indeed, propaganda is likely to be effective only if it is picked up and spread through personal contact. Cf. E. Katz and P. F. Lazarsfeld, *Personal Influence* (Glencoe, Ill., 1955), chaps. i–iii.

[9] E. Zacharias, *Secret Missions* (New York, 1946).

attempts to influence behavior in some relatively specific and preconceived manner. In psychological operations the purpose is to influence the behavior of the political decision-makers in a given foreign country. It is important to recognize that in some instances it is possible to affect the behavior of decision-makers by bringing about changes in the attitudes of the populace or select groups within the populace. If, for example, in a democracy public opinion swings sharply against a certain policy, political leaders may modify the policy even before the public has taken overt action to force a change. In this case a change in attitude in the populace contributes to a change in the action of leaders.

Finally, it should also be emphasized that the definition says nothing about the trueness or falseness or goodness or badness of propaganda. Propaganda merely tries to affect the manner in which people perceive the world and ascribe meaning to it. It may misrepresent the world, or it may try to bring peoples' perceptions more in line with reality.

While defining propaganda it is useful to make a distinction between two kinds of propaganda that may be called direct and indirect. Propaganda is direct when the immediate response desired by the propagandist is made explicit to the audience. Most advertising and domestic political propaganda is of this type—"Go to your nearest drug counter and buy *X* brand toothpaste"; "Vote for Joe Doaks." Institutional advertising is an example of indirect propaganda. The public is informed of the honesty and integrity of a certain firm in the hope that it will be encouraged to buy the products of the firm or to speak well of it to a friend who is in the market for such products. Much international propaganda is of this type. Straight news broadcasts to a foreign country are used to build up a reputation for honest reporting. American libraries are established overseas to help acquaint foreign peoples with American culture and values in the hope that this will make them look more sympathetically on American foreign policy and perhaps make their leaders do likewise. In indirect propaganda the behavioral or attitudinal response desired is not made explicit to the audience; however, if the propaganda is to be maximally effective, it must be clear in the mind of the propagandist.

Rumor.—A second technique of communication used in psychological operations is that of rumor. The definition of rumor used in this study is taken directly from Allport and Postman. These men define rumor as a "specific (or topical) proposition for belief, passed along from person to person, usually by word of mouth, without secure standards of evidence being present."[10] It can be seen immediately that rumor differs from propaganda. Initially anyway, it is limited in its audience. It is disseminated by personal contact and thus by the time it has reached a large audience it has been filtered through the minds of a great number of individuals. There is a certain amount of disagreement over just how reliable rumor is as a form of communication. Allport and Postman reported that typically some interesting changes take place in the process of transmission.[11] The source of the information is often lost or is changed. As the rumor travels, it tends to become more concise. Details, sometimes important details, are lost, but the substance that remains after repeated tellings tends to be sharpened and perhaps exaggerated. However, Schachter and Burdick have questioned this characterization and have performed an experiment indicating that, under certain conditions anyway, rumor can be a very reliable source of information.[12]

Some of the best examples of the use of rumor by a state are taken from experiences of the Second World War. Both sides deliberately planted rumors to be picked up by enemy agents and reported home. J. B. White in *The Big Lie* reports a number of these. One was the rumor planted by the British in various neutral capitals throughout the world where German agents operated that the British could defend their Channel coast by setting the sea on fire. Apparently, it was widely believed in Germany and led to experiments at Fecamp in Normandy with a barge protected by asbestos sheets. And one can imagine that it did not help the morale of German troops wait-

[10] G. Allport and L. Postman, *The Psychology of Rumor* (Cambridge, 1950), p. ix.

[11] *Ibid.*, p. 75.

[12] S. Schachter and H. Burdick, "A Field Experiment on Rumor Transmission and Distortion," *Journal of Abnormal and Social Psychology*, L, No. 3 (May, 1955), 363–69.

ing on the Channel coast to hear that their invasion fleet would have to pierce a wall of fire.[13]

The Germans also used rumor to try to affect the decision-making behavior of their opponents both before and during the war. White reports hearing a rumor in a Paris bar in 1938. A young German-American working in France for the Germans reported that a friend of his had been driving through Bavaria a few weeks earlier and had collided with a German tank on maneuvers. To his astonishment the tank fell apart and revealed itself to be nothing but a wooden frame covered with canvas and mounted on a Volkswagen chassis. Three weeks later White saw the same report in a well-known motoring magazine and eight weeks later it was repeated in the War Office Intelligence Summary (of course, it was highly classified). Apparently, the story was widely believed and used to support arguments that Britain did not have to rearm in 1938.[14] There is much evidence to support the proposition that rumor can be a very effective instrument to bring about changes in the way people perceive and interpret the world of events and objects.

Direct person-to-person communication.—The great upsurge in professional and academic interest in communications came after the advent of mass communication on a large scale. And consequently a large amount of the work that has been done in the field has been on the organization and impact of the mass media in modern societies. This should not blind one to the fact, however, that the most effective form of persuasive communication is face-to-face communication. The Soviets recognize this fact and the great amount of effort they devote to oral agitators is evidence of this recognition.

Direct person-to-person communication must be recognized as one of the techniques of psychological operations. The Communists have used this technique with effectiveness in many areas of the world. The cell organization of their party is an excellent device through which this technique can be used both in the Western world and among the peoples of Asia and Africa. A democratic nation can never exploit this method to

13 See J. B. White, *The Big Lie* (New York, 1955), pp. 1–10, for a discussion of the origins of this rumor and its effectiveness.

14 *Ibid.*, pp. 28–29.

the degree which the Soviets do. Certainly it would be impossible, owing to the totalitarian organization of Russian society, to have Western "agitators" operating there the way Communist agitators operate in the West. And even in other democratic countries the antipathy that would be created by the revelation that paid American agitators were operating might well more than nullify any benefits that might be gained.

But, as we can differentiate between direct and indirect propaganda, we can also discriminate between direct and indirect person-to-person communication. Many forms of student exchange programs and other forms of cultural intercourse are based on the assumption that foreigners will learn something about the United States through personal contacts while in this country, that they will take this understanding home with them, and that this will make many lasting "friends" for America who will interpret our policies sympathetically in years to come. And Americans abroad will by necessity bring something to foreign peoples with whom they have had prolonged contact and perhaps influence the manner in which these peoples look at the United States. While this very informal kind of person-to-person contact (with the object of the interchange never made explicit) may not be as effective as the direct agitation mentioned above, nevertheless it should not be overlooked as a technique of psychological operations.

Symbolic acts (or propaganda of the deed).—Meaning, of course, can be "transferred" through various types of action. Indeed, all foreign policy operations have a symbolic impact over and above their physical impact. But we are not imperialist enough to claim, therefore, that all foreign policy acts belong ultimately in the bag of techniques to be used in psychological operations. There are, however, instances when physical operations are important primarily for the meaning they convey rather than for the change in the material world which they bring about. Some of the Commando raids in France and the Low Countries in 1942 and 1943 were of primary importance not because of the physical damage that was done or because of the information gained about German defenses. But they suggested to the German High Command that an invasion was in the offing and thus tied down troops which could have been effectively used on the eastern or Mediterranean fronts. They also perhaps encouraged the captive populations to resist and harass their German masters. The Doolittle raid on Tokyo in 1942 is another example. The impact on meaning throughout the world was far greater in its implications than was the damage done to Tokyo.

There are also a number of peacetime examples of activities whose symbolic impact is of primary significance. The Moslem festival of *Id al Adha* is the time when thousands of Moslems journey to Mecca. In 1952 *Id al Adha* fell on Friday the twenty-ninth of August, but Mecca's gates were to close on the twenty-seventh. Less than a week before the gates closed there were more than 4,000 pilgrims stranded in Beirut—800 miles from the holy city. All had air tickets but the local airlines simply could not handle a fraction of that number before the deadline. After some amazingly fast and thoughtful work by the American legation in Beirut and by the departments of Defense and State in Washington, the Air Force rushed fourteen C-54's to Lebanon and began to airlift the pilgrims into a city only 40 miles from Mecca. Flying around the clock, they made it possible for all of the pilgrims to get into the holy city before the gates closed. The story of the airlift "magic carpet" was told in the Moslem press throughout the world, and the United States received favorable editorial treatment in a press that had been noted for its hostility.[15]

Although military aircraft were used, one could hardly argue that this was a military operation. It was an operation that indicated to Moslems in Turkey, Egypt, Iran, Lebanon, and Afghanistan that the United States, a foreign policy giant in the world, would keep an eye out for the "little guy" and was willing to help him if he got in a tight spot. It is difficult to conceive how much propaganda, rumor, and personal contact would have been necessary to get this message across in the absence of the airlift. Its impact was primarily symbolic.[16]

[15] See Daugherty and Janowitz, *Casebook*, pp. 337–42 for a complete report.

[16] It should be pointed out, however, that in our view of the nature of statecraft this operation becomes significant only if it affects the decision-makers' definition of the situation in a manner which is to achieve American foreign policy objectives. This particular operation might have had this effect. For example, a governmental leader in the Near East who was pursuing an anti-American policy might find it more difficult to get needed mass support in his country after the word of this operation was widely disseminated.

On the other hand, although the Berlin airlift had tremendously important symbolic effects throughout the world, one cannot say that its primary purpose was psychological. Once the Allied decision to stay in Berlin was made, the airlift became a necessity for circumventing the Soviet blockade.

In these kinds of physical operations it is extremely difficult to draw a clear line between a psychological operation and a military or economic one. But it is important to recognize that some desired changes in the apparent world can most effectively be provoked by a physical operation and that some of these techniques ought to be at the disposal of those responsible for psychological operation.

TACTICS

In addition to knowing the techniques available it is necessary to know something about the tactics that may be used. We find it useful to think of the tactics of SPO (Strategic Psychological Operations) in terms of two sets of dichotomies.

Deception-enlightenment.—These two tactics are best explained in relationship to the distinction between the material and the apparent worlds. Tactics of deception attempt to make the audience build up a psychological environment which differs from the material environment. In the terms of perceptual psychology, tactics of deception try to provoke non-veridical percepts, and tactics of enlightenment are attempts to bring a skewed perception in line with reality.[17]

There are many examples of the use of the tactics of deception. Much of Hitler's success in the late thirties was due to the fact that the official decision-makers in Britain and France defined the situation (particularly in regard to Hitler's goals) in a manner which deviated disastrously from reality.

There were many instances during World War II when deception was used effectively by both sides. We have already referred to the British rumor about setting the sea on fire and the German rumor of the collapsible tank. Deception is re-

[17] Much of the literature dealing with psychological operations associates tactics of enlightenment (strategy of truth) with democracies and tactics of deception (strategy of the big lie) with dictatorships. This is hardly justifiable. As was pointed out above, the nature of the originating society does place some limits on tactics, but they are not the rigid limits expressed by the formula "democracies speak (or should speak) the truth; dictators tell lies."

sponsible in some degree for the reputation of "evil" which tends to surround the psychological instrument. Indeed, among some people whose understanding of the dynamics of human behavior is meager and whose knowledge of world affairs is parochial, "psychological operations" are understandable only in what they refer to as "the dirty tricks department." All other aspects of the psychological instrument are apt to be scathingly referred to as "globaloney."

One of the most famous and successful uses of deception was the British "Operation Mincemeat"—the case of *The Man Who Never Was*. After the Allies had driven German and Italian military forces from North Africa, the next step in the Allied offensive was obvious to both them and the Germans. Before the Mediterranean could be available for shipping, Sicily had to be taken. The move was so obvious that the Germans could be expected to mass their defensive forces there and make the invasion, even if successful, extremely costly. The value of making the Germans think that the blow would come elsewhere was apparent. Operation Mincemeat was an attempt to do just this.

It involved releasing a corpse, dressed as a major of the Royal Marines, from a submarine off the coast of Spain. A "Most Secret" letter was planted in a dispatch case attached to the body. It appeared to be from General Sir Archibald Nye, the Vice-Chief of the Imperial General Staff to General Sir Harold Alexander, who commanded an army in Tunisia. The letter indicated that there would be a major Allied offensive against Greece. Other documents on the body were designed to lead the Germans to believe that the major was being flown into the Eastern Mediterranean on a plane that crashed off the coast of Spain.

The body was picked up by Spanish officials and the documents handed over temporarily to a German agent who made copies that were sent on to Berlin. After the war, captured documents proved that the German intelligence service believed the documents and convinced the High Command that the Allies would make their major attack in the Eastern Mediterranean. Military and naval forces were sent to Greece and held there even after the invasion of Sicily, because Hitler and the High Command were convinced that the attack on Sicily was diversionary.[18]

There were also a number of "black" propaganda operations undertaken by the Allies in World War II which employed the tactic of deception. "Operation Annie" and "Gustav Siegfried Eins" are examples of two radio stations that operated as if they were broadcasting from inside Germany. Their success was predicated on getting the Germans to believe that they were not Allied stations.[19]

There are three basic rules that must be followed if deception is to be an effective way of influencing behavior. First, the deception must be "reasonable." The success of Operation Mincemeat was due to the fact that it was entirely reasonable that a Royal Marine officer would be flying to North Africa with special messages in a plane that crashed, and that an Allied invasion in the Eastern Mediterranean was a reasonable if not the most likely move from Africa.

A second rule of deception is that there must be no simple way of checking what the facts in the case really are.

A third rule is that the use of deception should not discredit a source which may have valuable future potential. Deception is usually discovered eventually and the more successful the deception the more likely is the source to be discredited. In World War II the "black" Allied station, Operation Annie, was once used to direct a Nazi column into Allied hands. The deception was excellent, but it completely destroyed the future of the station. It is usually unwise to use a newspaper or radio station which has built up a large audience for purposes of deception.

The other part of the dichotomy, the tactic of enlightenment, is what is usually referred to as the "strategy of truth." It is a potent tactic, but hardly a magical one, as some of the American rhetoric would lead one to believe. Its potency derives from the fact that dire consequences may result for an individual if his world of meaning differs very greatly from the material world. The changes in the world in the last fifty years have increased the importance of the tactic of enlightenment.

18 E. Montagu, *The Man Who Never Was* (London, 1953).

19 H. Becker, in Daugherty and Janowitz, *Casebook*, pp. 672–77.

Until recent times most of the events and objects which were important in an individual's daily life were physically close to him. He could learn about them from firsthand experience; by and large, through his own devices the normal person could keep his apparent world relatively well adjusted to his material world. But one of the consequences of the physical "shrinking" of the world in the twentieth century, brought about by the technological revolution, is that many events and objects which may vitally affect the daily life of an individual are physically so remote that he has no opportunity to learn of them by direct encounter. He is far more dependent than previously on unknown and distant people to bring him knowledge that will enable him to avoid "bumping his nose" on an object or a fact that he did not perceive. The party that can provide accurate information therefore has a powerful instrument on hand to control behavior.

There are also rules which apply to the use of the tactic of enlightenment. Facts do not speak for themselves or at least identical facts may "say" different things to different people. A Hungarian refugee from the 1956 uprising reported to one of the authors that he was in a Budapest basement during the second Russian intervention waiting for the end when he heard a Western broadcast to Hungary that carried a verbatim account of the UN proceedings. When he heard the motion censuring the Soviets had been passed by a large majority, he thought of the Korean intervention, picked up his rifle and went back into the street to fight until he should be relieved by Western troops. The broadcast of the UN proceedings was as accurate a report as was humanly possible to present. But it meant something entirely different to a Hungarian Freedom Fighter from what it meant to a UN delegate or to the broadcaster transmitting it. Undoubtedly, it had behavioral consequences which were hardly desired or anticipated by those responsible for its broadcast to Hungary. To enlighten implies something more than "telling the truth." To use the truth to enlighten involves a thorough analysis of how the listener, given his beliefs and attitudes and his situation, will interpret the truth and translate it into action.

The example cited above is a case of misinterpreting the "facts." But there may also be instances in which the "facts" are clearly understood and yet the behavioral consequences are entirely different from those expected by the untrained observer. Festinger, Riecken, and Schachter have presented a convincing case study to indicate that when an individual has a strong belief to which he is committed by irrevocable action, the presentation of irrefutable evidence that his belief is wrong may not only leave him unshaken in that belief but lead to new and strong behavioral manifestations of intensified belief. The case study tells the story of a group of individuals who believed that they had made contact with superior beings visiting earth in flying saucers from the planet "Clarian." These superior beings passed on the information that much of the United States would be destroyed by floods on December 21. That day came and passed with no deluge. However, for a number of believers this unequivocal and irrefutable evidence that their prophecy was false did not lead to the abandonment of the belief but to a greater commitment to it and an increase in efforts to win new members to the cult.[20]

The authors list five conditions which must be met if such behavior is to result:

1. A belief must be held with deep conviction and it must have some relevance to action, that is, to what the believer does or how he behaves.
2. The person holding the belief must have committed himself to it; that is, for the sake of his belief, he must have taken some important action that is difficult to undo.
3. The belief must be sufficiently specific and sufficiently concerned with the real world so that events may unequivocally refute the belief.
4. Such undeniable disconfirmatory evidence must occur and must be recognized by the individual holding the belief. . . .
5. The individual believer must have social support. It is unlikely that one isolated believer could withstand the kind of disconfirming evidence specified.[21]

Are not these conditions met in some places in the Communist movement today? Are there not Communists in Moscow (and other places) who hold the official philosophy with deep conviction and have committed themselves to this belief

20 L. Festinger, H. Riecken, and S. Schachter, *When Prophecy Fails* (Minneapolis, 1956), *passim.*

21 *Ibid.*, p. 4.

through a lifetime of action? And have there not been events which have tended strongly to refute central portions of this belief? (For instance, the Hungarian uprising, or the miserable record of Soviet agricultural production for thirty years.) In this situation a propaganda which deals incessantly with irrefutable facts or instances when the Communist philosophy has been wrong may not destroy the "true believers'" belief but may actually strengthen it and encourage more proselytizing. Thus the behavioral consequences of broadcasting the truth in certain situations would be exactly opposite to those that may seem logical, if not inevitable.

It must be admitted that it is difficult to imagine how a philosophy like communism can be refuted as totally as a prophecy of cataclysm. But the hypothesis tested in the volume cited is derived from Leon Festinger's general theory of cognitive dissonance.[22] A relaxation of conditions 3 and 4 stated above would be more in line with the political example cited and would not be incompatible with the general theory. It may be more profitable, in directing the psychological instrument toward a Soviet Communist audience, to point out their leaders' distortion of the philosophy than to attack the philosophy itself. It may be more effective, in addressing, say, an Italian Communist group, to show the evils of the power-structure running from Khrushchev to Togliatti than it is to point out that communism itself has no real benefits to offer the Italian poor. The Festinger findings would seem to be supporting evidence for the theory that "broadcasting the truth" may have consequences diametrically opposed to those deliberately sought by the broadcasters.[23]

Terror and reassurance.—Terror has been used to control the behavior of men since the beginning of history. The dictators of the twentieth century have added new refinements.

[22] Cf. L. Festinger, *Theory of Cognitive Dissonance* (Evanston, Ill., 1957), pp. 24–28, 243–59.

[23] The theories of Festinger would also suggest that an individual's ideology may be the most difficult aspect of his system of beliefs and values to change. This suggestion is supported by much contemporary theory in anthropology and sociology. Indeed, when the only techniques available are those of psychological operations, it is difficult to conceive how one could have much success in attempting to change peoples' ideology.

The effectiveness of their use of terror to achieve political goals —both domestic and foreign—is a matter of historical record.

The effective use of terror is dependent on the use of coercion. Terror is essentially a threat, either implicit or explicit. If the threat is to be effective, it must be supported by physical violence—usually the more beastly, the more effective the terror. The coercion that is the most effective in instilling widespread terror is that which appears to its victims to be completely capricious. There should be no pattern in its application. If victims are chosen at random, all in the community may be terrorized; if the coercion follows a discernible pattern, only those who seem destined to be next will be maximally affected. In the Soviet Union in periods when terror is being accentuated as a tactic of control, no one really knows whether or not the next midnight knock on the door may not be at his room. A Stuka divebomber roaring out of the blue with its siren screaming picked its victims with no particular rhyme or reason.

There is evidence, however, that the use of terror (or more generally the use of fear-arousing appeals) may at times be an ineffective tactic. Hovland, Janis, and Kelley summarize a group of experiments in communication in which the effectiveness of the fear-arousing appeal was used by commenting:

> . . . when the goal of a persuasive communication is to create sustained preferences or attitudes, a relatively low degree of fear arousal will sometimes be the optimal level—that too strong a fear appeal can evoke some form of interference which reduces the effectiveness of the communication. The results [of the fear-arousing experiment] definitely contradict the assumption that as the dosage of fear-arousing stimuli in a mass communication is increased the audience will necessarily become more highly motivated to accept the reassuring recommendations contained in the communication. For communications of the sort used in the experiment, the *optimal dosage* appears to be far below the level of the strongest fear appeals that a communicator could use if he chose to do so.[24]

[24] C. I. Hovland, I. L. Janis, and H. H. Kelley, *Communication and Persuasion* (New Haven, 1953), p. 83. See pp. 56–96 for a complete discussion of the effectiveness of various kinds of fear-arousing appeals.

The question arises whether the use of terror as a tactic of psychological operations in the absence of overt hostilities is compatible with democratic values. Certainly, democracies have used terror. Some of the British bombing during the ill-fated Suez invasion of 1956 was undoubtedly an attempt to terrorize the Egyptian populace in the anticipation that it might turn against Colonel Nasser. Indeed the history of colonial operations is filled with the systematic use of terror to control native populations. But this is a despicable aspect of colonial history, and we would argue that the use of terror conflicts with democratic values and should not be used in time of peace. Actual war, of course, presents quite a different situation. There are instances in which terror might be the most "humanitarian" tactic available. The alternative might be dealing out wholesale death and destruction. It is feasible, anyway, that the threat to use hydrogen weapons in time of war might achieve some of the purposes of the actual use of the weapon and this possibility certainly always should be explored.

One must recognize that there are degrees of terror. Leaders in a democratic country will certainly feel restricted on the use of "unbridled" or "stark" terror in many situations. But they may very well attempt to "frighten" or "worry" foreign decision-makers. In a world where power is the only real arbiter between opposing giants, it would be suicidal, it seems to us, to renounce the use of fear or worry as a tactic of the psychological instrument of statecraft. Indeed, even such a moral and peace-loving man as President Eisenhower has employed it. The doctrine of "massive retaliation" is one example of its use. Another is the purposely vague policy in regard to what the United States would do in the event of a major Red Chinese assault on the offshore islands of Matsu and Quemoy. Obviously these doctrines are designed to cause worry and fear in any Sino-Soviet policy-maker contemplating armed aggression. A total renunciation of the psychological instrument's use of fear would make the maintenance of a military instrument a costly piece of nonsense.

The opposite of terror, reassurance, is also an important tactic that may be used in psychological operations. As man is sometimes influenced by fear, his behavior can also be affected if his fear is reduced. If the fear is "unreal," psychological operations in themselves can be effective in reducing it. But if the threat is real and intimate, reassurance can usually only be an effective tactic if it is combined with operations which remove or neutralize the source of the fear.

Of course, all four of these tactics can be used in the same campaign. Hitler's conduct of foreign affairs in the late 1930's provides an excellent example of the combinations of deception (as to motives), enlightenment (as to the power of the German military machine), terror (Poland and France), and reassurance (ambitions will be satisfied if the Sudetenland becomes German). Obviously, democratic governments cannot emulate Adolf Hitler unless they are willing to become the mirror image of that which they despise. But there can be an effective doctrine which specifies the integrated use of the technique and tactics listed above which will be compatible with a democratic philosophy.

A simple listing of techniques and tactics does not constitute a doctrine. Something additional is required. A doctrine is a series of propositions which state how various combinations of techniques and tactics should be used in various situations. In order to have a doctrine it is necessary to have some kind of a typology of situations. The doctrine would be a series of propositions of this nature: Given situation A, techniques of propaganda and rumor should be used employing largely tactics of deception and terror; given situation B, techniques of propaganda, face-to-face communication and symbolic acts should be used, employing largely the tactic of enlightenment. Situational analysis in the social sciences has not progressed very far, and when it comes to the specific kinds of situation typology which would be required to spell out a doctrine for psychological operations, there is nothing to go on. Although we must stop short of developing a doctrine, we have laid a basis for developing limited doctrines for specific situations.

DOMESTIC LIMITATIONS

It is rare indeed when a political instrument can be used to its full potential. Almost invariably certain limits are imposed by internal factors. Most obvious, of course, are the limits placed by the lack of manpower or material resources. A strategy is worse than useless if its successful implementation calls for people, tools, and skills which are simply not available. But there are also non-material limits which sometimes impose themselves. At times these limits may be related to such an integral part of the strategy-maker's culture that he fails to recognize them as variable limits. It is important for him to attempt to recognize them explicitly, because only by dragging them out for conscious examination will he learn whether or not the limits can be modified.

These domestic limitations on the uses of the psychological instrument can be divided into three closely related kinds: cultural, political, and material.

CULTURAL LIMITATIONS

The major value orientations of a society will certainly place certain limits on the use its government can make of all policy instruments. Earlier in this chapter reference was made to the fact that the tactics of deception and terror did such violence to democratic values that they should not be used in their ultimate form by the United States in the absence of overt hostilities. But democratic values place additional restrictions on psychological operations.

A modern democratic society is based on a philosophy which postulates man as a rational creature capable of making decisions that are in his own and in the public interest. It is a philosophy dedicated to the proposition that there is an inherent dignity in man that ought not be violated either by individuals or governments. There are times, as the Soviets have demonstrated, when it may be efficacious to engage in psychological operations which do violence to these principles. We would not argue that the democrat should dispense with his values while conducting foreign policy. Indeed, it is imperative that the tactics and techniques of a psychological campaign be carefully judged in the light of the dictates of the principles of democracy. And although in many instances the choice may be between two evils, it is well to look at alternatives from the point of view of democratic values.

But there are ways in which values can make unwarranted and dangerous intrusions into the area of policy-making. Perhaps the most serious is when value assumptions are given the status of empirical truths and become operating assumptions for strategy development. This tendency exists to an alarming degree in the United States. An excellent example came out in a press conference in the summer of 1957 when Secretary of State Dulles stated that American policy accepted as a "working hypothesis" the view that the type of despotism existing in the Soviet Union "will never prevail and that the kind of government which is responsive to the will of the people, which admits diversity and freedom of thought and expression is the government which has the future ahead of it. I am confident that [the trend toward freedom in the Communist world] is a basic truth. Certainly, it is an assumption that I think must be made by anybody who believes in the American Tradition."[25] Perhaps the Secretary was right in the eventual validity of his prophecy. But to develop a *working hypothesis* of this nature, not from empirical evidence, but from a belief in the American tradition, is to court grave trouble.

The implication of this statement is that a belief in the American tradition makes one accept the proposition that there is a basic trend for freedom and democracy throughout the world. But the image of the world around us does not give great support for this belief—at least not in the near future. One wonders if the Secretary was thinking of the Hungarian Revolt, while quite forgetting the completely effectiving smashing of this "diversity and freedom of thought and expression" by Soviet armed forces. How much diversity and freedom of thought and expression are allowed in South Africa, or Taiwan? What has happened to the brave new hopes for democracy in Indonesia, in Burma, in Pakistan?

The emergence of dictatorships or semidictatorships in many of the new lands recently released from a colonial past may be a necessary thing because their cultural, social, and economic base is not sufficiently developed to make political democracy viable. It is to be hoped that their "dictators," military and otherwise, will be beneficent and enlightened and that this is really only a developmental phase on the road to real democratic freedom. To date, however, there is little evidence to support the Secretary's optimism even in these non–Communist bloc nations. Ordinary citizens are unaware of what indications Mr. Dulles had that a "kind of government

25 *New York Times*, July 3, 1957.

which is responsive to the will of the people" will prevail behind the Iron Curtain. On the evidence publicly available one is tempted to be somewhat less sanguine. Moreover, it does not immediately become evident why a firm belief by Americans in the American tradition must lead them to believe the same blessings of liberty must inevitably fall on all peoples. It would perhaps be more reasonable, in the light of history, to expect the price of freedom to be high and to entail the wisest possible use of all the available instruments of statecraft. The world's apathy toward Germany between 1932 and 1939 was costly enough. How much more permanently disastrous would the results have been had all nations taken as "a working hypothesis" the "basic fact" that Hitler's government would never prevail? How disastrous may be the results if this working hypothesis is accepted now, vis-à-vis the Soviet Union and Red China, by the free world's decision-makers? The limits that such a hypothesis places on psychological operations are obvious. It is important for the strategy-maker to recognize them.

There are many instances in which value assumptions are looked upon as empirical truths. The American tendency to emphasize propaganda campaigns to wholesale rather than elite audiences might well be based on such a confusion.[26] The peculiar (often mystical) power that Americans seem to place on "truth" as a weapon in the Cold War also appears to be based not on rigorous, empirical analyses of the use of this tactic but on pure value orientations. It is important for the strategy-maker to be aware of these fallacies and to re-examine his operating assumptions to see whether they are empirically sound or merely transformations of some of the value assumptions of his own society.

There are also a whole host of cultural factors that place certain limits on psychological operations directed at a particular country. For example, the American emphasis on efficiency and practicality might be reflected in a propaganda campaign and seriously limit its effectiveness on a people who place greater emphasis on aesthetic, contemplative, or mystical concerns. Even to attempt to mention all the factors that must be

[26] Cf. Hans Speier, "International Political Communication: Elite *vs.* Mass," *World Politics*, IV (April, 1952), 305–17.

taken into consideration in regard to American operations in major areas of the world would go far beyond the scope of this book. Let it only be said that strategy-makers should beware of imputing their own values and the hypotheses of their own culture to their audiences.

There are other cultural factors which may inhibit the effectiveness of psychological operations. For example, the racial problem in the United States is such that it would be unwise to attempt to use certain psychological strategies directed at some audiences. These kinds of limitations should be obvious; further comment is unnecessary.

There is one other rather curious cultural limitation which drastically affects the American use of the psychological instrument of statecraft—the reluctance to use it! In 1955–56 the United States spent about $80,000,000 on overseas "information" activities. This is less than one four-hundredth of our military budget during that same year. Speier argues that there are many conflicting reasons for the skepticism concerning the value of psychological operations.

Apart from the opinion widespread in Protestant middle class civilization that "propaganda" is a form of deceit and therefor immoral, specialists often regard other instruments of policy—e.g. diplomacy, armaments, economic measures—both more respectable and more effective in the pursuit of foreign policy than the activities of journalists and radio script writers, or of the producers and distributors of motion pictures. Moreover, the technicians of the modern mass media are separated from the Foreign Service Personnel by differences in their respective professional codes of behavior—differences which give rise to reserve, impatience, irritation, misunderstanding, and sometimes distrust. By tradition and experience, the Foreign Service Personnel is inclined to observe time-honored ways of communicating with their peers in other capitals and to view communications to peoples at large as either wasteful or dangerous.[27]

POLITICAL LIMITATIONS

This analysis of the reasons for the reluctance of the United States to use the psychological instrument of statecraft introduces political limitations. And if one is looking for political limitations, one must look farther than the much-maligned

[27] *Ibid.*, pp. 306–7.

Foreign Service. Most people who have had experience with one of the U.S. information programs are extremely critical of the effect that Congress has on the organization and its program. There are many examples of unwarranted congressional interference. Leo Cherne's comments would be widely concurred in by those who have had experience with our overseas efforts in this field:

> The achievements of USIA are all the more remarkable in view of the handicaps under which it operates, the most brutal being Congressional interference. One incident will suffice—the recent insistence by members of the House Appropriations Committee that a certain American history textbook (one with a foreword by Charles Lindbergh and an introduction by Louis Bromfield, no less) be withdrawn from circulation in libraries overseas. Among the reasons: a photograph of "A little red schoolhouse, built 1750" might give the Communists the occasion to say this represents the American school system, although the photo was followed by pictures of a modern high school and of Harvard University; certain passages from "Walden" "might be damaging"; a graphic shot of a dust storm should be deleted because it was bad propaganda.[28]

Congress has also been somewhat capricious with the budget for the information program. By increasing it one year and decreasing it the next, it has forced the responsible agency to spend valuable time and energy merely enlarging or retrenching, rather than in conducting its overseas operations on the basis of long-range, coherent plans.

A government which has a separation of powers and a system of checks and balances has tremendous potential for friction and antagonism to develop between the branches. Edward W. Barrett, Assistant Secretary of State for Public Affairs from 1950 until 1952, feels that this antagonism is likely to be particularly apparent in the relations between any agency responsible for psychological operations and the Congress. Three of the reasons he gives are of particular importance:[29]

1. Because of the sensitive nature of psychological operations, their effectiveness suffers when Congress debates, for all to

[28] L. Cherne, "The Loin Cloth on the Rajah?" *Saturday Review*, September 17, 1955, pp. 13–14.

[29] E. W. Barrett, *Truth Is Our Weapon* (New York, 1952), pp. 254–55.

hear, the details of strategy and technique. It may tip off one's adversary on future actions and embarrass and insult a friendly country that has permitted large-scale American psychological operations within its borders.

2. Psychological operations are extremely complex. Congressmen rarely have the time to learn their intricacies and complexities. Therefore, ". . . full-scale Congressional debate is usually uninformed debate and often dangerous debate. . . ."
3. It is extremely difficult to estimate the effects of psychological operations. Their impact may be subtle; sometimes effects are not noticeable for years. This kind of activity is likely to be suspect particularly to the hard-headed businessmen who find their way into the appropriations committees. They tend to demand solid evidence of usefulness before they are willing to grant funds. Since the men responsible for psychological operations cannot provide a "cage filled with ex-communists converted by the Voice of America," adequate appropriations are often difficult to get.

MATERIAL LIMITATIONS

In addition to the political there are material limitations. Congress may, for example, provide enough funds for a tenfold increase in radio broadcasts to China, but the linguists may not be available. In time of rapid mobilization, psychological operations may be bidding against the military establishments for electronic equipment (and are quite likely to come up second best). There are also technical limitations. Soviet jamming has not completely blotted out VOA but it has made successful transmission spotty and costly in time, effort, and money. The Iron Curtain itself cuts out the use of many normal means of communication and often makes unusual means difficult. In various areas of the world newsprint is scarce and it is hard to have space allocated to the items fed to the press by American "propagandists." In some places radios are scarce and their range of reception very limited. Literacy, or rather the lack of it, is still a controlling factor in many countries. The successful strategy-planner must be aware of the limits that the availability of resources places on strategy development and implementation.

In this chapter the discussion has been limited specifically to a treatment of the requirements for developing a strategy for the employment of the psychological instrument of statecraft. But the psychological instrument should almost never be employed alone. It should be normally used with the other instruments to implement total strategies. This, however, raises the complex question of integrating the employment of the various instruments in tightly woven strategies. The need for integrated total strategies is obvious. The major obstacle to their development seems to be not so much one of understanding the need as it is of developing the kind of organizational machinery which enhances rather than inhibits this development. The problem of developing integrated total strategies involving the use of psychological operations is therefore a problem that will be treated in chapter iv on the organization of the agency responsible for psychological operations.

• CHAPTER III
The Intelligence Requirements

The rational choice of policy objectives is dependent on an estimate of the feasibility of various alternatives which in turn requires that at least the major outline of possible strategies be worked out. After objectives are selected, strategies must be developed in detail. The development of a strategy requires an estimate of the situation in which the strategy is to be employed. There are many aspects of any situation. Political, military, economic, sociological, and psychological data are necessary to an adequate estimate for the development of total strategies.

It is the responsibility of those in psychological operations to provide that intelligence which is peculiarly relevant to influencing the behavior of policy-makers in other states through the manipulation of the symbolic environment. But what kind of intelligence is relevant to this task?

Basically, there are three different, but closely interrelated, kinds of intelligence that are necessary for psychological operations. Early in the planning process the questions are asked: "Whose apparent world must be manipulated if we are to influence the behavior of foreign states in the desired manner?" and "What is likely to be the success of operations of various degrees of intensity?" These questions lead to a selection of audiences and to an estimation of the magnitude of the psychological operations required to achieve specified results. It provides what is known as *basic* intelligence. After audiences have been selected, the question is asked, "What should be said to the audiences that will first attract them to the communication stimuli and then tend to precipitate the proper responses?" The answer to this question requires *output* intelligence. Finally, it is necessary to inquire constantly into the adequacy of one's *basic* and *output* intelligence estimates as operations are being conducted in order to make the appropriate adjustments at the proper times. Intelligence that deals with the analysis of the effect of on-going or past operations is *feedback* or *evaluation.*

Basic Intelligence: The Selection of Audiences

There are many kinds of data that are necessary for the selection of audiences for psychological operations. Writers on this topic have repeated *ad nauseam* that data must be collected on audience attitudes, motivations, listening habits, the degree to which individuals might pass on what they hear, the economic and social position of individuals and groups and their position in the power structure of their respective countries, etc. But this kind of a statement is not too useful. If by some superhuman effort, all the data on all these subjects could be gathered, they would be of little use to the policy and operations planners. The planner needs information which has been analyzed, collated, weighed, and is presented in such a way that it is specifically pertinent to the planning process. It is this analysis and special presentation that differentiates intelligence from information. But how shall the vast quantity of sociological, economic, political, and psychological data be analyzed so that it will be useful to the planner?

In order to develop some relevant criteria for analysis and presentation, let us look at the planner's problem. The planner for psychological operations is engaged in the task of trying to develop a plan of action which will influence the behavior of foreign states in a preconceived manner through the manipulation of the foreign decision-makers' apparent worlds. In chapter ii it was pointed out that there are four basic approaches to this planning problem: (1) foreign decision-makers' perception of the external situation can be manipulated; (2) the external situation itself can be changed; (3) the foreign decision-makers' perception of the internal situation can be manipulated; (4) the internal situation itself can be changed. Each one of these approaches involves a number of potential audiences. If operations are aimed at influencing all these audiences, resources are likely to be so dissipated that operations are not likely to be successful. There must be some criteria by which audiences that will contribute maximally to the achievement of the objectives of the policy can be selected, and the data on potential audiences must be analyzed according to these criteria.

We propose that two criteria be used. They are (1) *susceptibility*—how susceptible the various potential audiences are to being influenced in the desired manner through psychological operations and (2) *cruciality*—how crucial the various potential audiences are in the policy-making process. These two concepts are simple enough in themselves and certainly easy to justify. The planner wants to select optimum combinations of audiences which he can influence and which in turn can influence the behavior of a foreign state in the desired manner. But simple though the concepts may be, the tasks of data collection and analysis necessary to apply them meaningfully to psychological operations are exceedingly complex. We will try to indicate the kinds of information and intellectual skills that are relevant to the determination of the susceptibility and cruciality of potential audiences.

THE CONCEPT OF SUSCEPTIBILITY

The planner knows the communications messages that he has available—he knows the response that he wants to achieve from his audiences. But there is no one-to-one relationship between the communication stimulus and the response. Exhortations to revolt, for example, may cause a few people to take up arms and others to think about it; they will more likely have no effect on many and even cause some to support their government with more fervor and patriotism than before. Man is not a cork adrift on a stormy ocean of propaganda, yielding passively to the strongest and most persistent currents. His response to communication messages is not to be understood solely or even largely through an analysis of the communication messages themselves. His reaction to communication messages must be looked at in terms of his whole pattern of needs, desires, fears, goals, frustrations, and achievements. The communication message goes through what we might call a "filtering and dyeing" process that may affect it greatly. Furthermore, it must be remembered that the recipient quite likely will be limited in the kind of behavioral responses he can make.

In order to determine the degree of susceptibility of various audiences it is necessary to understand the factors that may "filter and dye" the message and limit the responses of the audience. In the language of the social scientist these factors are known as intervening variables—they "intervene" between the stimulus and the response. But just what kind of factors "intervene" in the communication process? A specification of these intervening variables and the way in which they operate would be a *conceptualization* of or a *model* of the communication process. And data on which a susceptibility analysis of potential audiences rests are dependent on such a model. We do not feel that there is at present an adequate model of the communication process.[1] Nor do we propose to try to develop one here. It is more useful to draw on the great amount of research that has been done in communication to suggest the kinds of data that are relevant to the determination of susceptibility.

First, it is possible to distinguish between two logically different types of intervening variables—those which are aspects of the social structure of the potential audiences and those which are aspects of the personality characteristics of individuals in the audience. The first shall be called *societal* intervening variables and the second *psychological* intervening variables.[2]

Societal intervening variables.—The most obvious of the societal intervening variables are the channels of communication themselves. No audience could be considered susceptible if channels of communication to it are not open. And there

[1] See for examples of the work that has been done, T. R. Vallance, "Methodology in Propaganda Research," *Psychological Bulletin*, January, 1951, pp. 32–61; T. M. Newcomb, "Communicative Behavior," in R. Young (ed.), *Approaches to the Study of Politics* (Evanston, Ill., 1958), pp. 244–65; and B. L. Smith, "International and Intercultural Communication: A Theoretical Model" (Ph.D. Dissertation, University of Chicago, 1957).

[2] This distinction is taken from Vallance, p. 32.

are a whole host of factors that may block the channels. The channels may not be open because a totalitarian regime makes listening to foreign broadcasts a serious crime (political factors), or the potential audience may not be able to afford radio receivers (economic factors), or interference may make the signal inaudible in certain places (technological factors).[3] Different social groups in a society may also rely on and react to the various media of communication in quite different ways.

The work done by the Russian Research Center at Harvard University on communication patterns in the Soviet Union is suggestive of the kind of research that is necessary for an adequate consideration of the societal intervening variables.[4]

In Table 1 the exposure of different social groups to various kinds of communications media is reported. It clearly indicates that workers and peasants will be exceedingly difficult to contact with the media and techniques available to U.S. psychological operations in the Soviet Union.

It is also important to be aware of the fact that the same media may perform different functions for different social groups. Table 2 indicates how various social groups differ in the importance they attribute to different media.

The tables reveal two characteristics about communication patterns in the Soviet Union that are important in any assessment of the susceptibility of various potential audiences. First, we should draw attention to the importance of word-of-mouth communication. It is almost as important as the newspaper and far more important than any other media, including radio. Second, there are great differences in communications patterns among the various social groups. The urban intelligentsia and the peasants represent the extremes, and they will be discussed in more detail. But the fact that these two polar groups are discussed should not obscure the fact that the phenomena referred to tend to vary continuously between the poles.

[3] Adapted from E. Katz and P. F. Lazarsfeld, *Personal Influence* (Glencoe, Ill., 1955), p. 21.

[4] See A. Inkeles and R. A. Bauer, *The Soviet Citizen: Daily Life in a Totalitarian Society* (Cambridge, Mass., 1959), pp. 159–88; and R. A. Bauer and D. Gleicher, "Word of Mouth Communication in the Soviet Union" (Research Memorandum No. 15, Air Research and Development Command, Human Resources Research Institute, Maxwell Air Force Base, Alabama [August, 1953]).

TABLE 1*

Social Class Patterns of Exposure to Communications Media

Type of Media Exposure	Intelligentsia	White-Collar Employees	Skilled Workers	Ordinary Workers	Collective-Farm Peasants
Official Mass Media Exposure:					
High	52%	38%	22%	7%	3%
Medium	45	47	38	32	21
Low	3	15	40	61	76
Personalized Media Exposure:					
High	14	7	6	3	2
Medium	33	18	9	6	6
Low	53	75	85	91	92
Aesthetic Media Exposure:					
High and Medium	63	44	35	18	8
Low	37	56	65	82	92
Covert Media Exposure:					
High	20	11	7	2	1
Medium	26	25	16	10	12
Low	54	64	77	88	87
Total number of respondents	642	679	282	494	387

Mass Official: newspapers and magazines, books and Soviet radio
Aesthetic Official: movies and theater
Personalized Official: lectures, agitation meetings, and other official sources
Covert: discussion with friends, rumor, foreign radio

* A. Inkeles and R. A. Bauer, *The Soviet Citizen: Daily Life in a Totalitarian Society* (Cambridge, Mass.: Harvard University Press, 1959), p. 167.

TABLE 2*

Information Sources Most Important for Various Social Groups

Media	Intelligentsia	White-Collar Employees	Skilled Workers	Ordinary Workers	Collective-Farm Peasants
Newspapers	47%	59%	30%	35%	18%
Word-of-mouth	34	23	26	43	60
Radio	9	9	22	14	18
Meetings	2	4	0	2	0
Other	8	5	22	6	4
Total number of respondents	100	56	27	49	27

* A. Inkeles and R. A. Bauer, *The Soviet Citizen: Daily Life in a Totalitarian Society* (Cambridge, Mass.: Harvard University Press, 1959), p. 168.

1. The Peasantry

The heavy dependence on oral communication as a source of information does not differentiate the Russian peasant from peasants in other parts of the world. Lower educational levels, geographic isolation, and the technical inadequacy of the formal media in rural areas in non-industrial societies contribute to that dependence. But the fact that the Russian peasantry is among the most disaffected social groups in the Soviet system increases its indifference to the formal media.

Keeping the overall importance of word-of-mouth communication to the peasants in mind, three analytically distinguishable functions of this system for the peasant can be cited. First, word-of-mouth communication provides the peasant with much legitimate and immediately useful information. News of agricultural policy, of the arrival of goods at local stores, and of new administrators comes to the peasant through informal channels, as well as information about national and international events. The sources for this information tend to be the few people in the community who read a newspaper or may have a radio, or people who are able to travel from one community to another. Second, the word-of-mouth network provides the peasant with much illicit and otherwise unavailable information—news of economic failures, of bad living conditions, of indications of hostility to the regime, etc. In addition to giving the peasant a more complete picture of life and conditions in the Soviet Union, the word-of-mouth system functioning in this regard may also give the peasant a certain amount of emotional satisfaction. It is interesting to note that the peasantry is the only group for which frequency of contact with the unofficial communications system correlates positively with the expression of anti-Soviet attitudes.

As the system operates in rural areas, there tends to be considerable "distance" between the original source and the individual peasant who finally receives the information. As is typical in the transfer of information, the more intermediaries involved, the more likely it is that the original information will be distorted. This fact coupled with the additional fact that the peasant is not particularly well equipped to evaluate critically the truthfulness and reliability of information makes him especially vulnerable to rumors of low plausibility.

2. The Urban Intelligentsia

The professionals, contrasted to the peasantry, are very actively involved in communication activity—official and unofficial. Their involvement in the unofficial network can be identified by four characteristics. First, the intelligentsia, while preparing reports, analyzing production figures, and having occasional access to the foreign press, are in a good position to observe the facts and conditions in the official reports and see commentaries on these facts. This not only makes them aware of information "gaps," but enables them more or less accurately to locate these gaps. Second, because of his firsthand information and the circles in which he moves, the highly educated urbanite is in a position to fill in these gaps in the official media by word-of-mouth contacts. Third, because of their position in the Soviet hierarchy, the intelligentsia are particularly accessible scapegoats. Thus it is to their advantage to be in a position in which they can get advance knowledge of policy changes or at least of how to gain access to the information that may enable them to anticipate such changes. It might be said that they are "under pressure" to maintain contact with reliable unofficial sources. A fourth characteristic of the intelligentsia's participation in the unofficial system is a relief from sheer boredom. The Soviet press and radio are noted for their ability to repeat endlessly and monotonously the same propaganda line in the same phrases year in and year out. A kind of information fatigue sets in and the intelligentsia "escape" to the unofficial network in order to find some relief.

But we must not forget that the intelligentsia's prime source of information is the official media—particularly the newspapers. In contrast to the peasantry for whom the unofficial network tended to be a substitute for the official sources, the intelligentsia use it to supplement and correct the official media. It provides them with information not available officially and provides a check on the accuracy of that which is available. The reliance on word-of-mouth indicates a high degree of reliance on one's sources of information. The interviews reveal that, whereas only 59 per cent of the peasantry feel that rumors were more reliable than the press, 90 per cent of the intelligentsia so indicated.[5]

We are not here using this study for the substantive data it records on the Soviet Union, but because in its treatment of communication in the Soviet Union at a given period, it provides a number of valuable clues about the societal variables that intervene between the communication stimulus and the response. First, the analyst must be aware of the possibility of using a number of different channels of communication. Second, different kinds of distortion may occur in different channels of communication among different groups. It would be very difficult, for example, to predict the content of a communication message a Russian peasant heard over the Voice of America after it had been retold several times. Third, different channels of communication may perform quite different functions for different groups. It would be important for planners to know about these differences so that messages could be fed into channels in relationship to the functions these channels play for the audience. Fourth, different groups may be differentially susceptible to different tactics and techniques simply because of their different reactions to the channels employed.

On the basis of these data on the media of communication and their functions among various groups in the Soviet Union, the urban intelligentsia would have to be ranked as the most susceptible potential audience and the peasant the least susceptible. The intelligentsia can be contacted by the use of the techniques of psychological operations, while the peasant will be very difficult to reach. Although the peasant may be ranked more susceptible on the basis of other factors involved in a susceptibility analysis, the absence of channels of communication open to him will keep him in the ranks of the least susceptible.[6]

The distinction made in this report of communication patterns in the Soviet Union between mass media and word-of-mouth communication should not obscure the fact that they

[5] Inkeles and Bauer, p. 176.

[6] The fact that the peasant will also have to be rated very low on a cruciality ranking will make him a very low priority target, his disaffection with the Soviet regime notwithstanding.

are often involved together as part of the same basic pattern of communication. In fact, communication research in the past two decades has made it very clear that the mass media have their impact as their messages are carried into various groups and interpreted by "opinion leaders." In other words the mass communication process seems to be a "two-step" process. There are certain individuals that mediate between small, face-to-face groups and the "outside world." Part of a susceptibility analysis should be concerned with studying the primary group structure of a potential audience in order to understand what kinds of individuals tend to play the role of opinion leader and what kinds of messages they will receive favorably and encourage the circulation of throughout the group.[7]

There is another way in which the small primary group structure of a potential audience is relevant to a susceptibility analysis. There is a very impressive amount of evidence to indicate that an individual's attitudes which may appear to be private are in fact developed, supported, and maintained in conjunction with a small number of others with whom he is motivated to interact.[8] As psychological operations may try to change the attitudes of various audiences it is important to know how various primary groups will resist or support the desired changes.

The evidence for these statements about the role of small group structure in the communication process comes largely from work done in the United States. However, it is supported by work done elsewhere. For example, in Lerner's book on the modernization of traditional societies, there is described a village in Turkey where not many years ago the only radio belonged to the chief. The people in the village who heard the radio directly were the friends of the chief, the kind of programs they heard were determined by the chief, and the information received was discussed under the watchful eye of

[7] See Katz and Lazarsfeld, pp. 1–132, for an excellent review of a great deal of small group research and its relationship to studies of the communication process.

[8] See for a review of relevant research and theory: Edward Shils, "The Study of the Primary Group," in D. Lerner and H. Lasswell (eds.), *The Policy Sciences* (Stanford, Calif., 1951), pp. 44–69; and Dorwin Cartwright and Alvin Zander, *Group Dynamics: Research and Theory* (Evanston, Ill., 1953), *passim.*

the chief. It is clear from the whole series of reports on the Middle East that primary groups are very significant to an understanding of the process of mass communication.[9]

There is one other societal intervening variable that must be mentioned. The social structuring of an audience will place limits on the responses which will be made to communication messages. This is particularly obvious in a totalitarian country, but it operates to some degree in all societies. The probabilities that certain kinds of responses can or cannot be made are very important aspects of any susceptibility analysis.[10]

These rather random examples of the societal intervening variables are perhaps misleading in their apparent simplicity. But they are merely indications of the many variables within societies and within groups in societies that are elements of the social structure which may affect communications. The factors are not independent of one another or of the psychological intervening variables. An adequate susceptibility analysis will be partially dependent on propositions, not only about the factors themselves, but upon their interrelationships.

Psychological intervening variables.—There are situations in which the attitudes of an individual may completely alter the meaning of a communication message. The customer in a used car lot who hears from the salesman, "This beauty was owned by an old-maid school teacher who never drove faster than twenty," may not have his views on the car altered, but the message may say to him, "That shyster salesman thinks I'm a sucker." The attitudes of the audience are important to the communication process for two reasons. First, they affect the manner in which the communication stimuli are perceived. Second, the attitudes themselves may have to be affected first if the communication messages are to precipitate the desired behavior on the part of the audience.

The psychological intervening variables and the perception of the communication message.—It has long been recognized that there is no one-to-one relationship between events and objects and the manner in which events and objects are per-

[9] Daniel Lerner, *The Passing of Traditional Society* (Glencoe, Ill., 1958), *passim.*

[10] See H. C. Kelman, "Attitude Change as a Function of Response Restriction," *Human Relations,* VI (August, 1953), 185–214.

ceived. Differences and discrepancies have long intrigued psychologists. It would not be an exaggeration to say that a whole school of psychology, the Gestalt, developed on the basis of answers to the question, "Why do things appear as they do?" However, it was not until the 1940's that psychologists widely recognized what a significant effect an individual's needs, beliefs, and values could have on the way in which he perceived the world. The dramatic experiments of Ames and Cantril and their associates, mentioned earlier, indicate that great discrepancies could occur between an object or situation and a person's perception of them. Sherif demonstrated how personality traits could affect the manner in which the social environment was perceived. Bruner and Goodman in the well-known experiment in 1947 indicated more specifically how needs and values could affect perception in a crucial manner. A group of children, some from wealthy and some from poor families, was asked to adjust a diaphragm so that the amount of light showing through was the size of a number of coins ranging in value from a dime to a half-dollar. They were also asked to select cardboard disks that were equal to the size of various coins. Although all the children tended to overestimate the size of the coins (the greater the value of the coin, the greater the overestimation), the poor children tended to overestimate the size more than the rich children did. These results were interpreted to indicate that need partially determines value assigned to objects and together they may affect perception.[11]

The point is that perception is a selective process. An individual's needs and values and attitudes can affect perception in three major ways. First, there is a great deal of omission that takes place. Certain objects which are a part of the total picture may never be "seen." Second, perception may also involve some supplementing. There may be some unconscious "rounding out of the picture"—some additions so that what is seen "makes sense" in terms of a person's own view of the world. Third, and most important, a great deal of structuring

[11] J. S. Bruner and C. C. Goodman, "Value and Need as Organizing Factors in Perception," *Journal of Abnormal and Social Psychology*, LXII (1947), 33–44. A large number of other experiments have been performed corroborating the results obtained by Brunner and Goodman. For a review see F. H. Allport, *Theories of Perception and the Concept of Structure* (New York, 1955), pp. 256–466.

will take place. Certain elements will be highlighted, others played down. The listener to a communication message will perceive it in such a way that it makes sense to him in terms of his past experiences and his present activities.[12]

Unfortunately, the experimental and survey research that has been done on the perception of communication stimuli has not been explicitly related to the vast amount of experimental and theoretical work that has been done in perception. Research has been done on discrete problems, with little attempt to select the problems in relationship to any perception theory or to develop research designs which would facilitate making generalizations from the results that would have significance beyond the narrow limits of the problem under consideration. However, some communications research does indicate the manner in which an individual's system of beliefs and values affect his perception of communication stimuli.

The most conclusive results deal with the effect of attitudes toward the communicator on the perception of the communication message.[13] Most of this work revolves around what is referred to as the "credibility" of the communicator or the analysis of "prestige effects." Hovland, Janis, and Kelley distinguish two important factors: (1) the extent to which a communicator is perceived to be a source of valid assertions (his *expertise*) and (2) the degree of confidence in the communicator's attempt to communicate what he considers to be valid assertions (his trustworthiness). In the experimental studies which demonstrate the importance of these two factors, characterization of the communicator is varied experimentally; in the survey studies, differences in the audience's perception of the communicator are analyzed in relationship to effects that the communications have had.

For example, in one experimental study college students were given passages which were represented as excerpts from newspaper and magazine articles on the feasibility of developing an atomic submarine. One group was told the article was by J. Robert Oppenheimer, another that it was a translation

12 Adapted from T. M. Newcomb, *Social Psychology* (New York, 1950), pp. 90–96.

13 This literature is reviewed in C. I. Hovland, I. L. Janis, and H. H. Kelley, *Communication and Persuasion* (New Haven, 1953).

from *Pravda.* The same article had significantly different effects on the changing of opinions.

Robert Merton's famous survey study on Kate Smith's bond-selling marathon clearly demonstrates that one of the most important factors in the success of this campaign was that the audience perceived Kate Smith as a very sincere person.

Hovland, Janis, and Kelley sum up the research results on the effect of estimates of source-credibility on the acceptance of communication messages as follows:

1. Communications attributed to low credibility sources tended to be considered more biased and unfair in presentation than identical ones attributed to high credibility sources.
2. High credibility sources had a substantially greater immediate effect on the audience's opinions than low credibility sources.
3. The effects of opinion were not the results of differences in the amount of attention or comprehension. . . .
4. The positive effect of high credibility sources and the negative effect of the low credibility sources tended to disappear after a period of several weeks.[14]

Although these conclusions stem from work on American audiences, they strongly suggest that the potential audiences' images of what they believe to be the source of a communication message will have a considerable effect on how they will react to the message. Audiences that believe the communicator of the message is knowledgeable and trustworthy are, other things being equal, going to be more susceptible. Research on this factor is an essential part of any susceptibility analysis.

The audiences' views on the communication channels of messages also may affect the response. The data on communication in the Soviet Union presented above clearly indicate a case in which estimates of reliability are based to a significant degree on the image of the communication channel. This is another psychological intervening variable that must be taken into consideration.

As important as the attitudes of the audience toward the communicator and toward the channels of communication is the relationship between the beliefs and values held by the audience and the substance of the communication message. The

[14] *Ibid.*, pp. 269–70.

person who agrees essentially with the content of a communication message is likely to perceive it differently from the person who disagrees with it. Unfortunately, little work has been done on how the congruence of an individual's values with those of the communication message might affect his perception of the message. One of the studies done on American troops during the Second World War does indicate the relationship between attitudes held on a given subject and the most effective form of argument to influence attitudes about it. The results indicate that a person who disagreed with a given position was most influenced by an argument that stated both sides of the case and then proceeded to accentuate in one way or another one side. A person who basically agreed, however, could be made secure in his belief by a one-sided argument.[15]

Not only opposition and support must be taken into consideration; indifference is an equally important factor. One of the important characteristics of perception that must be taken into consideration is the omission of certain features. Objects and events taking place in a person's surroundings are more likely not to be perceived if they are not salient to his belief-value system. For the normal human being the things that are perceived and acted upon are the things he is interested in. Estimates of susceptibility must be partially based on estimates of how important a given issue is to the individuals in the potential audience.

There seems to be an implied assumption among propagandists that attitudes are somehow "softer" than concrete behavior—that it is easier to change a man's attitudes and through changes in attitude precipitate changes in behavior than it is to influence behavior directly. This is a tenuous assumption at the best. A man's basic belief-value system is usually very resistant to change, particularly by use of the techniques available to psychological operations. In many situations it might well be easier to influence behavior by attempting to change a man's cognitions rather than by attempting to modify his belief-value system. In other words, instead of trying to influence behavior by trying to change motives, one might try to affect the way a man perceives that his goals can be achieved. This

15 See C. I. Hovland, A. A. Lumsdaine, and F. D. Sheffield, *Experiments on Mass Communication* (Princeton, 1949), pp. 201–13.

kind of approach usually is more useful than others in operations against foreign leadership groups.

In chapter ii reference was made to the suggestion that if in World War II, after the flight of Rudolf Hess to England, broadcasts to Germany had hinted at the possible defection of some important figures in the Nazi party, they might have contributed to suspicion and intrigue—possibly even to a purge—which would have weakened the German war effort. No attempt would have been made to change any of the top leaders' goals. The suggestions of possible defection, however, might conceivably have altered some top leaders' ideas about steps that must be taken to achieve their goals.

Another example may be more to the point. The history of the Red Army indicates that political leaders have often been concerned about the loyalty of officers and men. There have been experiments with "political commissars" at virtually all ranks to check on loyalty. These political officers were not trained primarily as military experts, and in periods when their influence was the greatest the fighting effectiveness of the Soviet armed forces seemed to decline. In the event of open hostilities against the Soviet Union a concerted, well-publicized drive to induce the soldier to surrender might cause the leaders in the Kremlin to increase the power of the political officers in the armed forces, thus reducing the likelihood of battlefield decisions being made on the best military grounds.

One might call this the "judo" approach to psychological operations. In any social system there are some factors at work which tend to limit efficiency and effectiveness. The "judo" approach would involve seeking out these factors and attempting to enhance them, or, using another metaphor, attempting to make a vicious circle more vicious.

This approach, of course, is not limited to use in conjunction with miltary operations. It would seem very difficult to try to change in one way or another the attitudes of the urban citizens in the Soviet Union toward communism and the Soviet system through broadcasting a few hours a day over the Voice of America. However, the strong and unsatisfied desires for more consumer goods might be whetted and strengthened by the right kind of programs and these desires, as they become stronger and more clearly articulated, might help place some limits on Soviet economic war potential.

These comments on the "judo" approach provide a good closing note for our discussion of the criterion of susceptibility. The purpose of a susceptibility analysis is to discover and rank all potential audiences on the basis of the probability of manipulating their behavior using the techniques and tactics of psychological operations. A whole host of interrelated factors will be relevant to the analysis. They might be listed in summary form as follows:

1. Societal Intervening Variables
   a) *Channels of communication.*—One must know about the channels of communication that are available to reach certain audiences. It would be necessary to have data on literacy, radio-set ownership and distribution, and the patterns of face-to-face contacts which are important in the transmission of messages. It is also necessary to have estimates of the kind and direction of distortion apt to occur.
   b) *Limits to responses.*—There will always be social limits to the responses that can be made to communications messages. The nature and stability of these limits, the probabilities that certain responses will be made and the circumstances under which certain responses will be made should be known.
2. Psychological Intervening Variables
   The attitudes of people in the potential audiences which will affect the perception of the message must be estimated. Attitudes toward the perceived source and the various media are, for example, essential data. It is also necessary to study the attitudes, needs, and values that are important in the motivational and cognitive patterns of people in the potential audiences which will be important in understanding the way in which certain kinds of messages will lead to certain kinds of behavior responses.

It is not difficult to understand that an estimate of the susceptibility of various audiences to psychological operations is a necessary objective of a basic intelligence estimate. But the kind of analysis required for this estimate is exceedingly complex.

THE CONCEPT OF CRUCIALITY

Susceptibility cannot be the only criterion for the selection of audiences. The most susceptible audience would be a poor focus for psychological operations if its reaction has no effect whatsoever on the behavior of the state. Any foreign policy strategy is a plan of action to attempt to influence the behavior of foreign states. Audiences, therefore, must be selected in relationship to their ability to contribute to the kind of state behavior which the strategy is trying to elicit. A ranking of the potential audiences on the basis of the degree to which they can influence the behavior of the state is what we call a *cruciality ranking.*

If the behavior of that abstraction *the state* is conceived to be the behavior of those who authoritatively act in its name—the official decision-makers—then an estimate of the cruciality of various potential audiences must be based upon some concept of the decision-making process. There are several approaches to the analysis of the decision-making process that could be adapted for this purpose. For purposes of discussing in more detail the concept of cruciality, we will adapt the one developed by R. C. Snyder, H. Bruck, and B. M. Sapin.[16]

Figure 1 draws attention to the important factors that must be taken into consideration in an analysis of the decision-making process. Let us look at it as a diagrammatic portrayal of the state whose behavior is to be influenced. The box labeled *E* (action) represents the official acts of the decision-makers—the behavior of state "X." Usually, the individuals and groups which are most crucial in the determination of that behavior are those in box *D*—the decision-makers themselves. There are a few cases in which it is possible for the psychological instrument to work directly on top-level decision-makers. Operation Mincemeat is an excellent example.[17] However, these instances are rare. Most psychological operations are directed at groups and individuals who are not a part of the official decision-making apparatus. The line from *B* to *D* on the diagram indicates an interrelationship between the social setting and the decision-making process. Some of the social groups (*B5*) may have an impact (direct or indirect) on the decision-making process. A ranking of the groups according to their influence on the decision-making process is the

[16] See Snyder *et al., Decision-making.*

[17] See chap. ii, pp. 34–35.

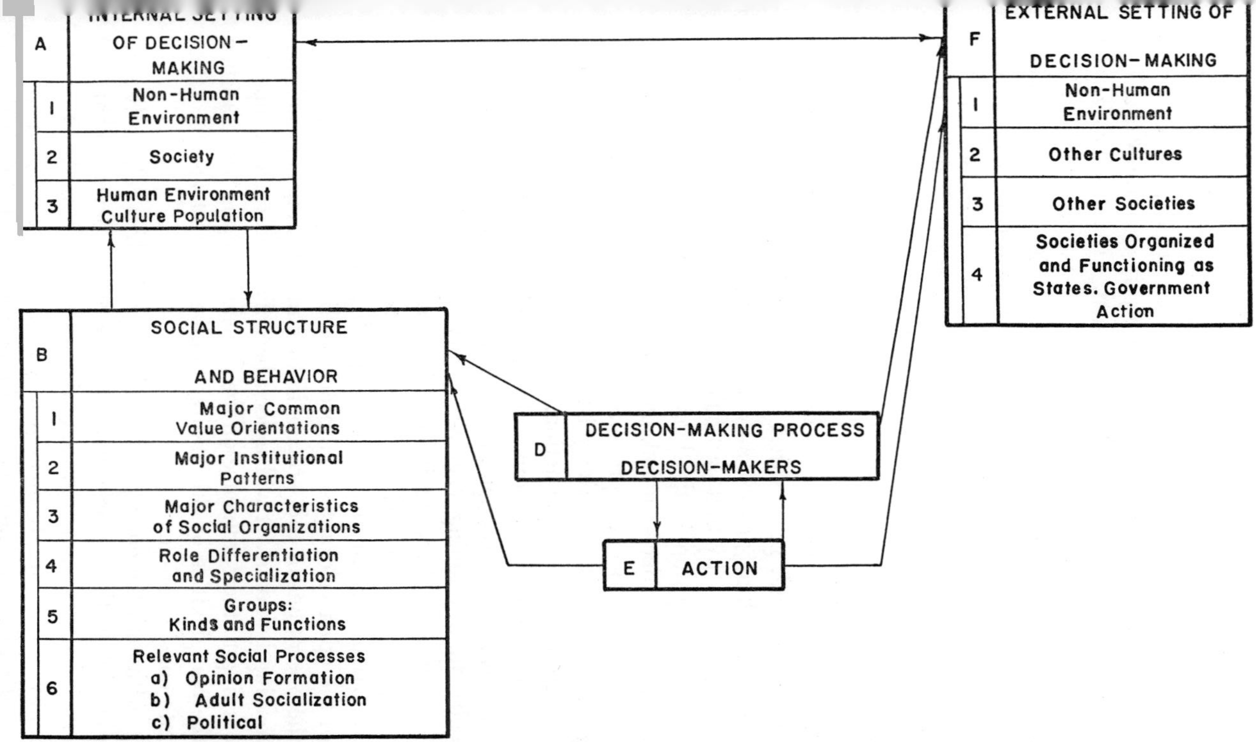

FIG. 1.—State "X" as actor in a situation. The situation is comprised of a combination of selectively relevant factors in the external and internal setting as interpreted by the decision-makers.

From R. C. Snyder, H. W. Bruck, and B. Sapin, *Decision-making as an Approach to the Study of International Politics* (Organizational Behavior Section, Princeton University, June, 1954).

major aim of a cruciality analysis of a potential audience country.

The development of a complete cruciality ranking would be exceedingly complex and difficult, even if all possible information were available, which is rarely if ever the case. Two special difficulties should be mentioned. First, various groups may have influence in relationship to certain particular issues and not to others. For instance, in the United States the Committee for Economic Development might conceivably have considerable influence on the kind of tariff law that is passed but very little in relationship to an executive decision to negotiate with a foreign power for an exchange-of-persons agreement. This difference derives primarily from one of two factors. First, a particular group may be interested in only certain issues and not use its influence to affect the outcome of others. Second, groups that have "access" to one part of the decision-making process may have influence over matters handled primarily by that particular part but not over others. For instance, a group in the United States which has its primary access to the decision-making process through the legislature will be able to exert pressure on matters dealt with by Congress but may be relatively impotent in regard to matters that require only executive action.

The implication of this point is obvious—not one, but several cruciality rankings are required. The second major difficulty in developing cruciality rankings is that groups may have either a direct or indirect influence on the decision-making process (or both). For instance, certain groups may have a particular ability to influence the "opinion process" (or one of the other relevant social processes listed in *B6*, Fig. 1) and the operations of this process in turn might influence decision-making behavior. It will indeed be difficult to rank groups according to their influence on the decision-making process when the complexities of this indirect influence must also be taken into consideration.

Although the concept of cruciality in itself is relatively simple, the further conceptual and theoretical development and empirical research and analysis necessary to make it maximally useful are tremendous. And in the discussion above, mention has only been made of cruciality in relation to the objective of influencing the behavior of existing decision-makers. If the objective is to overthrow these decision-makers, the system which must be analyzed is different. There are two possibilities. In states that have established institutions for the peaceful change of decision-makers (democratic states) an analysis of the process involved in the change is necessary in order to determine how this process can be influenced. Where peaceful change is not possible, some kind of revolution or coup d'état is necessary and an analysis of the process involved and of the groups crucial to this process is required. Again the complexity is great, but analysis is possible and without it decisions on audience selection are going to be more or less hit-or-miss.

Output Intelligence

In chapter ii we told of the young Hungarian Freedom Fighter who, as the Russians moved their tanks into Budapest, went to hide in a basement to wait until the occupation of the city was completed. But he heard a broadcast relay the debates from the United Nations wherein the Soviet Union was censured. He immediately thought of the UN intervention in Korea, picked up his rifle and went back into the streets to fight until UN forces arrived. Obviously, the Americans who were responsible for the relay broadcast did not want it to precipitate that kind of behavior.

This little account is a tragic way of saying that oftentimes words do not mean the same thing to different people. It is from this basic fact that the need for output intelligence arises. Once the planners have decided what kind of behavioral responses they want to elicit and audiences have been selected on the basis of an analysis of their cruciality and susceptibility, it is necessary to devise the communication messages to accomplish the desired results. Output intelligence provides the data which are necessary to design communications which will (1) attract the selected audience and (2) precipitate the proper response.

The major problem of output intelligence is not so much one of analysis as one of collection. There is no good substitute for close personal contact with the audience, for this is how one can determine the kinds of words and phrases that will best carry a message home. Sometimes this close personal contact is difficult. Herz reports how intelligence personnel during the Second World War got output intelligence for tactical operations: "We ate, slept and drank with prisoners. Many nights I was awakened by members of our crews dragging in deserters, who sat on my bedroll, dripping the waters of the river Scorff as they told me the latest Winchell dope on what went on inside the fortress."[18]

But many times such personal contact is not possible, and costly substitutes are necessary. For example, Radio Free Europe has an elaborate information-gathering service. Thirty-five radio stations behind the Iron Curtain are monitored on a regular basis. The aural monitoring averages between 45,000 and 48,000 words daily; the mechanical (Hellschreiber or radioteletype) between 125,000 and 150,000. The library at RFE receives 250 satellite publications which are carefully analyzed. In addition there are a number of field offices ranging from Istanbul to Stockholm to London. Correspondents from these field offices interview refugees and legitimate travelers from behind the Iron Curtain and Westerners who have returned from trips. The correspondents send information thus gained into RFE's European headquarters in Munich in a steady stream. In addition to providing much material that forms the basis of news broadcasts and commentaries, this information is analyzed to provide output intelligence.[19]

Experience seems to indicate that output intelligence calls for quite different intellectual skills than does basic intelligence. Ingenuity, intuition, spontaneity, and sympathetic understanding of, or empathy with, the audience seem to be more important than the capacity for cold, detached analysis. For this reason it is necessary to have different personnel perform the functions even though the basic information which they analyze comes from the same sources.

The analysis problems in regard to output intelligence are those inherent in all intelligence work. The major problems of this kind of intelligence seem to be not so much what kind of special analysis is needed but how to get the basic material

18 Daugherty and Janowitz, *Casebook*, p. 386.

19 R. T. Holt, *Radio Free Europe* (Minneapolis, 1958), pp. 97–119.

that is needed. However, this problem goes beyond the limits of this book.[20]

Evaluation

There are few problems of psychological operations as difficult as assessing impact. As there is no one-to-one relationship between the communication message and the response, it is extremely difficult to isolate what aspects of a response can be attributed to a communication stimulus from those that come from other sources. The situation is complicated by the fact that the psychological instrument will usually be employed in combination with the other instruments of statecraft. Many times to isolate the effect of one from the effect of another is a task of immeasurable proportions.

In the last decade there has been some excellent research which has begun to isolate and measure the impact of mass communication.[21] But the problem is far more complex when dealing with international communications and psychological operations. As Davison and George point out:

> The study of communication is sufficiently complicated if we confine it to the domestic scene. When we turn our attention to international political communication, where the "who" is a complicated propaganda apparatus in one culture, the "whom" is an amorphous audience in another culture, and the purposes and circumstances are bound up with all the intricacies of international relations, then it is clear that we are not yet qualified to undertake a *systematic* study of international communication.[22]

Overcoming the complexities of designing research to measure the effect of psychological operations is a problem that cannot be treated in detail in this book.[23] However, one point should be made. Feedback should be directed not only at

20 See Daugherty and Janowitz, *Casebook*, pp. 485–97; and D. Lerner, *Sykewar: Psychological Warfare against Germany, D-Day to VE-Day* (New York, 1949), pp. 94–130.

21 See Katz and Lazarsfeld, pp. 137–381, for a report of a most sophisticated attempt to isolate and measure the communication process.

22 W. P. Davison and A. L. George, "An Outline for the Study of International Communication," *Public Opinion Quarterly*, Winter, 1952/53, p. 502.

23 For a report of an intensive effort to study the effectiveness of Western broadcasts beamed behind the Iron Curtain, see Holt, pp. 120–44.

measuring the effect of psychological operations, it should also be designed to evaluate the accuracy of the original basic intelligence estimates of the susceptibility and cruciality of the audiences selected and of the adequacy of the output intelligence. This should not be looked upon as something different from measuring effect, because if the audiences were indeed susceptible and the output intelligence accurate, there should be changes in attitude and/or behavior; if the audiences were crucial, these changes should have some effect on the state's behavior. This formulation of the problem does not make the task of evaluation any easier, but it does help to provide a focus for the efforts that are undertaken.

• CHAPTER IV
Organization and Personnel

It is perhaps a truism to say that an organization should be so constituted that it is capable of effectively carrying out the functions assigned to it. This may sound a little like saying that a ball should be capable of being rolled or that a stove should give heat. It is, however, not quite so simple or so simple-minded as it sounds. And it is sometimes useful to repeat the truth in a truism. For political or sociological or economic reasons many organizations, "psychological" organizations among them, have not been designed in such a way that accomplishment of their proposed function is easy, or sometimes even possible. Sometimes such agencies fade away, but sometimes they do not; they hang on for years, being kept alive with periodic injections of funds or new blood in the hope that this will enable them to "do the job somehow." More often, happily, if there is a recognized need for the function to be served, a reorganization, or a series of reorganizations, takes place. The constant reorganizing which takes place within government agencies is a reflection of conscious effort to achieve an effective means of performing a function believed to be necessary. Frequently the change takes place because events have modified the scope of the function to be served. Sometimes, however, reorganization results from a revised understanding of the function itself.

The Background: The Relation of Function to Organization

The first step then, in logically advocating an organizational structure to perform a given mission, is to be certain that all concerned are in accord as to the meaning and the scope of the functions to be performed. The definition of the function must be understood and the necessity for the function itself must be accepted. In chapters i and ii it was argued that an objective of statecraft is to evoke desired political decisions on the part of the leaders of some foreign nation or nations. It was further argued that effective pursuit of this objective in the twentieth century calls for the integrated use of all four instruments of statecraft. The argument was developed, in both chapters i and ii, that these four instruments may be equal, or that in any given strategy applied to any given situation any one, including the psychological one, may be dominant. It is also apparent that any one of the instruments, but particularly the diplomatic or psychological, can be the initiator of strategy to be applied to a given situation.

Both chapters, but particularly chapter ii, indicated some ways in which the psychological instrument of foreign policy can be used. If the argument to date has been accepted, then it may be also accepted that there is a need for an organization specifically responsible for using this instrument in an effective manner. This may be considered the function of the organization on the highest level of abstraction. On a lower level, there are certain internal, or intra-agency functions which are necessary prerequisites to, or bases for, the performance of the higher-level function. Three of the most difficult of these, policy-making, planning, and intelligence, were discussed at some length in the preceding chapters.

This chapter will attempt to outline a type of organization capable of performing its functions on both of the levels of generalization mentioned above. It will not attempt to draw up a complete blueprint but will emphasize particularly those segments of the organization which are considered vital or which are most apt to encounter difficult sledding for any of a variety of reasons. Concurrently with the discussion of the organization's structure, and again particularly in connection with the key elements, some reference will be made to the personal attributes and qualifications of the personnel necessary to fill the jobs in question.

SIMILARITY OF FUNCTION IN PEACE AND WAR

Generally speaking, an effective organization for the use of the psychological instrument of foreign policy can, with relatively minor adjustments, serve the nation in peace or war, or in that in-between area of neither "peace" nor "war" which we know as the Cold War and which stretches before us into the future. Our present governmental arrangements draw too precise a line between peace and war. The military services each acknowledge a wartime "psychological warfare" role. Each admits, on paper, a mission of supporting the "Cold War objectives" of the United States. What evidence we have indicates that even the Army, which has in the past more consistently than the others planned for "psychological warfare," has since 1957 let this function suffer almost complete collapse. On the other side of the ledger, the United States Information Agency (USIA) disclaims any "wartime" role for itself and has therefore declined to cooperate with the services in planning for joint efforts in the event of a general war. We would argue that USIA's response is a further manifestation (perhaps forced on them) of Washington's shunning of real responsibility for proper and full use of the psychological instrument. Obviously the psychological aspect of military operations and maneuvers increases in periods of overt hostilities, but the existence, nature, and deployment of armed forces have psychological aspects at *all* times. Whatever the nature of the times, then, the full potential of the psychological instrument can only be realized if some organization controls its use on such a basis that long-range and coordinated plans can be made. In the course of this chapter where appropriate, the preparations for, and the modifications necessary to, use of this instrument during a period of actual warfare will be mentioned.

THE IMPORTANCE OF STATUS TO EFFECTIVENESS

To conduct strategic psychological operations successfully, the organization responsible must be properly placed in the overall government structure. This is to say that it must have a prestige status equal to that of the agencies employing the other major instruments of statecraft. Its position must be such that lateral access to other agencies involved in foreign policy is rapid, flexible, and non-interruptable. It must be such that vertical access to the National Security Council, the Cabinet, and the President is not only *possible,* but *easy.* As was mentioned earlier, the structure of the U.S. government is complex in the extreme. Given the size of this country and its power position in the modern world, there is small chance that it can move in any direction other than toward further growth and further complexity. Under these conditions of a multiplicity of authority and responsibility it becomes increasingly difficult to make integrated use of all the major instruments of statecraft. If the *status* of the agency conducting strategic psychological operations is on a lower level of importance than, say, State or Defense, it simply will not be able fully to exploit the potential of the psychological instrument. Perhaps this argument can be most clearly presented by using the SPO agency's chief as a "model," for after all, his position-status and the attention paid him by other agency chiefs is but a reflection of the position-status and the perceived importance of the agency itself. Conversely, if the head of the agency is accepted and respected on the highest governmental levels, the importance of his agency and its function are insured.

The present organizational structure of those elements of the U.S. government chiefly concerned with foreign policy—those agencies wielding the four instruments of foreign policy—places the agency principally concerned with the psychological instrument (or that part of it we appear willing to recognize and use) at a distinct status disadvantage vis-à-vis the agencies principally concerned with the other three instruments of statecraft. The chart (Fig. 2) may not perhaps directly indicate that the SPO agency is on a *lower* level than State and Defense, but, by taking USIA out of the hierarchical parallel with those agencies and setting it apart from the "command" lines for operations—and placing it in an "advisory" position—it immediately signalizes to "official" Washington (always very alert to such subtle distinctions) that this is indeed an agency of lesser importance. The present governmental organization chart (for foreign policy matters at the top levels) looks somewhat as shown in Figure 2.

Several comments about the chart need to be made. It should be noted that State and Defense are coequal (the Cabinet level) and that the U.S. Information Agency is significantly not in that category. Three other important facts emerge from a status-relationship study of the chart. (1) USIA is "under" the NSC (National Security Council), which in turn, for all practical purposes, is "under" the guidance of a special assistant to the President and not directly under the President

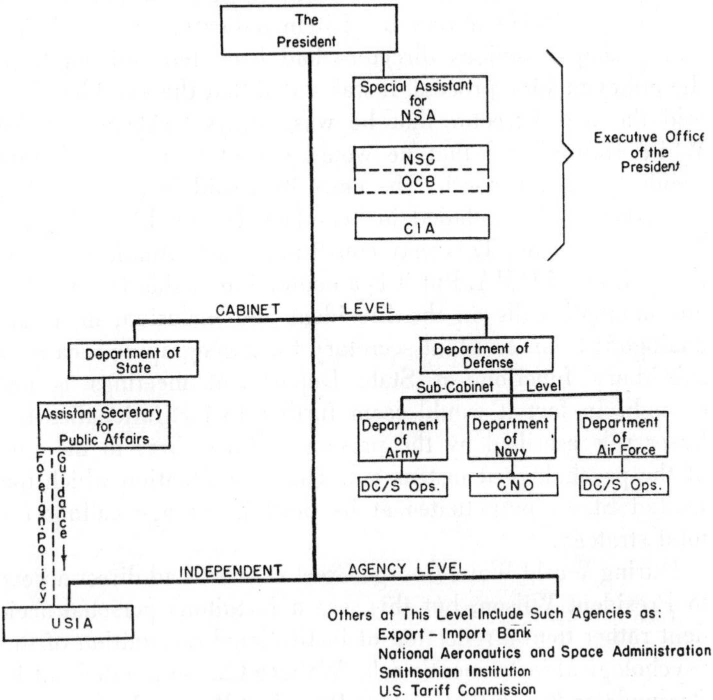

FIG. 2.—Schematic position of psychological operations—U.S. government, 1960.

himself, as are State and Defense, the wielders, between them, of the major portions of the other three instruments of statecraft. (2) USIA is placed on a different level and in a more remote position (vis-à-vis NSC or the President) than the Central Intelligence Agency (CIA). It is axiomatic that CIA, an intelligence agency, does not formulate policy but performs a requisite supporting function. By the peculiar sort of syllogistic reasoning in status-conscious Washington, USIA, if it does anything at all in relation to NSC, does so less importantly than CIA. (3) USIA, instead of itself being on the policy-making level, receives "policy guidance" from a relatively lesser official of the State Department. It would also appear not insignificant that that official is the head of what amounts to the "public relations" function of the State Department.

Shortly after George V. Allen assumed the position of Director of USIA there appeared in the *New York Times* (November 22, 1957) a very brief item indicating an admission that perhaps previous directors had been too isolated from the policy-making process. It was stated that the President had told the new Director that he was always "welcome at the White House" and that he would expect to meet with him regularly once a month. The news item said further that Mr. Allen was to be included in Secretary Dulles' biweekly conferences at State. This may constitute a new trend, a sort of recognition of USIA. But it is a contention of this volume that once-a-month calls on the President and inclusion, in a role analogous to an assistant secretary for a geographic area or a subsidiary function, at State Department meetings is not enough. In fact it would seem further to institutionalize the lesser role ascribed by the present policy-makers to the role of the psychological instrument. This is a situation which the United States perpetuates at its peril in an age calling for total strategy.

During World War I George Creel in effect had direct access to President Wilson, but this was a fortuitous personal accident rather than a reflection of institutional recognition of the psychological instrument's role. Wallace Carroll pointed out in *Persuade or Perish* that neither President Roosevelt nor Secretary Hull ever really understood the role of the Office of War Information. There is no evidence that either Presidents Truman or Eisenhower has understood the instrument's role any better.

The head of the Strategic Psychological Operations (SPO) agency should be a regular attendant at Cabinet and National Security Council meetings. He should hear policy being discussed while it is still in the formulative stage. He should be in both an actual physical locus[1] and in a figurative power position to be continually able to listen and to contribute to

[1] He cannot have his headquarters in New York or Battle Creek.

these discussions. To consult, or to be consulted by, one or more Cabinet officers or their staff members, after the fact, is not a satisfactory solution. It is unsatisfactory on several counts.

First, after a Cabinet or NSC meeting has broken up and the participants have dispersed to their various departments, it is extremely difficult to recapture the spirit of joint approach to the problem or policy just discussed. At the meeting, at the moment when all present are grappling with the same problem, in one another's presence and subject to one another's comments, pressures, and even prejudices, it would be easy for the head of SPO to express his views, air his fears, or outline hopeful modifications. There are very real physical difficulties in seeing, or even telephoning to, a half-dozen or more extremely busy men located all over Washington. Moreover, no amount of later consultation on a bilateral basis can be as effective as open discussion around a table when all concerned are present and in a frame of mind to face the problem at hand. To see each man individually, hours or days later, is to annoy each on a matter he feels has already taken his time; it is also a way to waste much of the SPO chief's time. Certainly there are Cabinet or NSC meetings at which matters are left undecided, because a major protagonist declines to divulge his full position for his own or his agency's tactical reasons, and sometimes this lack of final agreement is for the express purpose of allowing for further study and consultation. But even from these, many members will go with their minds tending toward some suggested solution. Even in these cases an excluded head of SPO would find it difficult to achieve the degree of receptivity to his ideas that he would have found at the meeting itself. But in this day and age, because of the press of time and the variety of problems to be dealt with (and no one who has not fairly recently been exposed to the upper levels there can possibly appreciate the press of both time and the multiplicity of problems on Washington officials) most meetings will adjourn only after a general consensus on broad issues has been reached. This means that the non-participant is faced with the task of attempting to change or modify decisions already reached or to sway minds already made up.

Second, the exclusion of the chief of SPO from Cabinet and

NSC meetings is unsatisfactory because his contribution (its importance is assumed, and it is assumed that he has seen and consulted with the interested participants) may be too late for the onrush of important world events. The United States must be geared to act and to react fast to all sorts of circumstances. Korea, Formosa, Suez, Hungary, Lebanon, Cuba, the Congo, are but a few of the names which made headlines in the mid-twentieth century. Each of these and many other such headlines signified a situation calling for the speedy integration of every foreign policy instrument available. Decisions had to be made fast, and policies and strategies had to be developed, made explicit, and implemented in the most rapid fashion. Each policy decision vitally affected the U.S. position, not only in the area named, but all over the world. Each decision involved strategic psychological considerations. Each demanded that what informal and bilateral consultation could take place should be undertaken quickly and on an unplanned-for basis. There was no time for altering decisions or changing the course of action taken by the President and his principal advisers after they had met to determine what the decisions and courses should be. In each case the world was waiting for a statement of the U.S. position; it would not wait long and still accord us the role of leadership.

The third reason why the exclusion of SPO's chief from a seat at the policy-making table is not satisfactory is that such exclusion automatically demotes the SPO agency *and function* from the level of importance on which it should be maintained in this age of instantaneous mass communications methods. To be effective in Washington, not only must the chief of SPO know that his is an important job and that his advice must be given, but all agency heads must also recognize the importance of SPO and seek its counsel in all matters affecting the U.S. position in the world. To make the head of SPO an "outsider" among the senior Washington officials concerned with foreign policy is to mark him as someone of lesser rank and therefore of lesser importance. This is not conscious snobbery; it is merely a fact of life and a recognition, perhaps unconscious, that in government circles, importance, like water, seeks its own level.

There are those who would argue, not entirely without justification, that our senior councils are already too large and unwieldy to act as fast as circumstances seem increasingly to demand. If this is so, then anyone advocating further enlargement of these policy and "command" bodies must make a strong justification for his advocacy. In this case the requirement is something less than this, for the fact is that the head of USIA often, if not usually, is present at NSC meetings. But the point needs repeating, that at present he sits there as a "junior member," as a "guest," not as an equal to the members designated by law and Executive Orders. It may suffice here to stress only two functions which the head of SPO can fulfil at these top-level meetings. And his special *expertise* in these will not, in the realities of Washington, have the weight which it must have accorded it unless he speaks as an equal to those with whom he sits. One deals with the question of feasibility as discussed in chapter ii. Whether or not an objective and the alternatives to achieve it are feasible may well depend on the apparent worlds of the people and nations affected. The chief of SPO should therefore be in a status-position to bring to bear fully at NSC meetings the related and important findings of that special sort of intelligence outlined in the preceding chapter. The other function this man is peculiarly and perhaps uniquely capable of performing bears on the psychological implications which even the Jackson Committee conceded inhere in *all* foreign policy activities. If any use is to be made of beneficial implications, and if disadvantageous ones are to be avoided, then the activities must be prejudged in this light. All the evidence points to the damaging psychological implications of the Roosevelt-Churchill "Unconditional Surrender" statement. Had a trained and capable SPO man been available and had his counsel been listened to, that statement *might* not have been made, and the war in Europe *might* have been shorter. This, of course, is an entirely "iffy" question, as Roosevelt used to say, but it serves as an example of a crucial moment in history in which, properly used, the psychological instrument could have been of great value to top policy-makers and to the nations they served.

Assuming that others have answered the question of "why" any given policy objective, the chief of SPO should be in a position to contribute, equally with State and Defense, to the question of "how." And while they can help answer the question of "at what cost" in men and dollars, he as well, and sometimes better than, the Secretary of State can really answer the question of at what cost in prestige, in respect, in adherence, or in alienation insofar as other peoples and nations are concerned.

THE IMPORTANCE OF THE CHOICE OF MAN TO HEAD SPO

Since we have been discussing the governmental place of the agency in terms of its chief official, it seems appropriate here to say something of what kind of a man he should be. At the outset one must admit frankly that he will meet much opposition to his agency because its function is so thoroughly misunderstood and mistrusted. He must therefore be a man who is not easily turned aside or ignored or circumvented. He must be a determined man whose determination is based on a full understanding of his agency's function and the nation's need for the exercise of the psychological tool of "total strategy." But since knowledge and determination are not enough to win acceptance for his agency and its function, he must also possess great tact and persuasive powers. Determination, tact, and persuasiveness are personal characteristics which many other key men in the SPO agency must also have in good measure, particularly those whose jobs will bring them into contact with members of Congress and upper-echelon men in the older, more established departments of government.

It goes without saying that any man should know his job. What may not be so clear is that in most of the policy, planning, and operational positions of this agency officers must have a very keen understanding of politics—international and national. International, for the obvious reason that their function is part and parcel of the forces working in the international political arena. A knowledge and understanding of the national politics of the "audience" countries is equally obviously a necessary tool in the kit of the SPO operators. The need for a knowledge of U.S. politics is less obvious, but that does not mean it is less essential. Chapter ii dealt with the capabilities of the "acting" country to use the psychological instrument. Some uses are, and some are not, politically feasible. But in addition to that need for knowing American politics, the SPO operator has another; a part of his function is to explain the "why" of U.S. policies and actions. This cannot be done successfully—that is, in such a manner that the foreign audience will understand and believe—unless the expositor himself fully understands what he is attempting to explain. Part of the SPO function, it has been argued, is also to help shape U.S. policy and to help guide and time its actions. These are the job primarily of the agency head, but his key men all the way down the line must be constantly aware of these responsibilities and assist their superiors, through the chain of command, to pass appropriate ideas and recommendations upward through the agency. None of them can properly discharge these duties unless they are keenly aware of both the historic and current political forces and trends in the United States.

In the normal course of events all Cabinet officers and major agency heads are aware of at least current political events and have manifested some interest in them, if nothing more than financial. Otherwise it is most unlikely they would have attracted enough presidential attention to be asked to serve. Traditionally, however, there have been certain posts which were considered suitable for "partisan" politicians and others which were somewhat above the purely partisan level. The position of Postmaster General and Secretary of State, respectively, are the two most apparent extremes. Partly, it is assumed, Presidents pick men as Secretaries of State whose interests and background far transcend those of party politics. Partly it is no doubt a result of the fact that a post of such grave national responsibility breeds within the incumbent a personal sense of responsibility to the nation above that of party. A "national," as differentiated from a "party," foreign policy has happily almost always emerged, no matter what party came into power as a result of no matter what campaign oratory. This is not to say that the "outs" and the "ins" have no differences over what foreign policy should be, or that there are no differences in actual foreign policy between the parties; it is merely to say that in our history we have been fortunate that our Presidents and our Secretaries of State have usually acted out of sincere concern for the nation and have tended to feel that purely national partisan matters "stop at the water's edge."

The National Security Council (NSC) establishes very broad policy in matters of national security necessarily stated in very general terms, and almost invariably carrying a very high security classification. The Operations Coordinating Board (OCB), a subagency of the NSC, takes these broad policy statements and breaks them down into specific goals to be attained. There may or may not be time limits set on the attainment; some may be considered as continuing or constant. The OCB further designates, for each goal, certain tasks to be accomplished by the individual departments and agencies of government. Often a task is to be performed by more than one department or agency. In this case, one is assigned primary responsibility and the other agencies concerned are given "support" roles. The OCB, then, is the spot in the governmental structure where, on paper at least, the process of integrating the foreign policy instruments takes place. The Director of USIA is a member of the OCB, where integration supposedly takes place, but not of the NSC, where policy is formulated.

There would seem to be little question that, if he is to help establish broad policy, the chief of SPO should be a member of the NSC as well as of the Cabinet. It would be even more important, if a choice had to be made, that he be on the NSC except for the fact that a member of the latter who is not also a member of the Cabinet would suffer from not having heard the background discussion which takes place in the Cabinet. He also would suffer from a certain lack of status within the National Security Council, where his colleagues responsible for the other instruments of statecraft are Cabinet members.

Obviously, he or his deputy must be a member of the OCB, since this is the point of foreign policy integration. There are, however, distinct drawbacks in the structure of this agency as presently constituted. Not the interested Cabinet members (State and Defense), but their deputies, are members of the OCB. Furthermore the assignment of tasks is now a fairly cut-and-dried process in which the agencies are assigned responsibility for tasks they are already performing in one way or another anyhow. The final drawback is that the OCB assignment of tasks is binding only insofar as the agencies concerned accept them; the OCB has no directive authority. This means that, although the Under Secretary has accepted the task, unless he, within his own agency's chain of command, orders and follows up its execution, no one feels compelled to make any special effort to implement it. The present system for this process of integrating is better than none, but it will never be fully effective unless one of two things takes place. Either all the OCB members must be so imbued with the importance of the OCB function and the necessity for following its assignments in spirit as well as in letter that each orders his own department's compliance, or the OCB must be given some authority for supervision and follow-up. The latter would present organizational difficulties and a storm of protest from the departments who jealously guard their own prerogatives. The former solution is obviously not easy either, but it may be the only feasible one. Its success or failure would depend very largely on the ability of some OCB member to convince the others of such a necessity. This job would appear logically to fall on the shoulders of the SPO chief, who, of all of them, has fewer other interests and problems and who because of the wide scope of his instrument's use should most thoroughly appreciate the need for integration of it with all the others.

In summary, it can be said that the head of Strategic Psychological Operations must be a man of broad vision and imagination. He must have a knowledge of national and international politics; he must have a knowledge of the world and its variety of problems. He must be devoted to the nation and to the philosophy behind the democratic process. He must have a broad cultural background. This man must be tactfully determined and diplomatically forceful. He must be articulate and persuasive. He will almost certainly not be a businessman (since so few have the inclination or opportunity to develop the necessary qualifications), but he must have achieved sufficient prominence in academic, governmental, or other professional pursuits to be listened to by his government colleagues. He is not an easy man to find.

The Present Structure for U.S. Use of the Psychological Instrument

The question of whether the SPO organization should be a part of the State Department or should be a separate agency follows very closely the argument which has gone on for years about the place of the foreign "information" agency. This argument really started in the early days of World War II when the United States first entered the arena of conscious, planned use of propaganda as an instrument of foreign policy. President Roosevelt "settled" the matter by Executive Order 9312 in May, 1943, by "placing full responsibility in the Office of War Information for foreign propaganda activities involving dissemination of information."[2] The President's action was really a wartime expedient taken at a time when most people, including the President himself,[3] did not fully appreciate the psychological instrument and certainly had little inkling of its future need in peacetime. This expedient merely put off the day of final decision; it took into the Executive Office an agency toward whose function the State Department had already manifested both lack of understanding and high distaste.[4] This lack of appreciation of the new instrument of statecraft was one of several reasons for the wartime decline of State's prestige and power-position among agencies having overseas interests.

It was not until the end of 1944 that the State Department really acknowledged the need for postwar "informational" activities.[5] It was then that Archibald MacLeish was given the newly created post of Assistant Secretary of State for Public and Cultural Relations. But this certainly was not an answer to full exploitation of the psychological instrument in the conduct of foreign affairs. The major emphasis was on domestic relations; State still saw its overseas task as one of furnishing "information" about America and Americans to foreign peoples.[6]

WIDESPREAD LACK OF UNDERSTANDING OF SPO

In further evidence of American high-level lack of understanding of the use of, or indeed of the necessity for, the psycho-

[2] U.S. Bureau of the Budget, *The United States at War* (Washington, 1946), p. 230.

[3] Wallace Carroll, *Persuade or Perish* (Boston, 1948), p. 7.

[4] J. P. Warburg, *Unwritten Treaty* (New York, 1946), pp. 72–73.

[5] U.S. Congress, House Hearings, *National War Agencies Appropriation Bill for 1945*, 78 Cong., 2 sess., 1944. Cf. various testimony, pp. 8–9, 36, 169, 333–37.

[6] *Ibid.* pp. 6, 179–83.

logical instrument of statecraft, almost before the sound of shooting had died out, President Truman abolished the OWI and created the Interim International Information Service (IIIS) within the State Department. This agency was directed to terminate its affairs on December 31, 1945.[7] The President thus gave State a four-month period to readjust itself to the assigned task of foreign information. In September, William Benton succeeded MacLeish as Assistant Secretary, and the department created the Office of International Information and Cultural Affairs (OIC). This office then proceeded to assume all the informational and cultural functions of State, including those of the temporary IIIS.

By the end of 1945, the inherited functions of the Office of War Information and of the Coordinator of Inter-American Affairs (CIAA) had very nearly ceased to exist. The "information" program was now little more than an augmented service of the embassies; the cultural and scientific cooperation program was confined to the Latin-American area, and the Voice of America had sharply curtailed its broadcasting.[8] It should not be thought that State alone was lacking in vision in this field; as a matter of fact, under Assistant Secretary Benton, new horizons began to be perceived, spurred on by the events of the then just beginning Cold War. In the Fulbright Act[9] the Congress gave some evidence of an awakening interest in the general field of cultural exchange as a part of U.S. foreign relations. This act extended the possibility of cultural and professional exchanges beyond the area of Latin America. The State Department was feeling its way and attempting to arrive at workable solutions for a use of the psychological instrument which remained largely not understood. Mr. Benton, in 1947, proposed that Congress establish a public corporation to carry on the informational and cultural cooperation work.[10] Congress was not receptive to this idea. It was not until the summer of 1947, when a great many congressional committees

7 Executive Order 9608, August 31, 1945.

8 Cf. C. A. H. Thomson, *Overseas Information Service of the United States Government* (Washington, 1948), p. 191; and Department of State, *A New Instrument of U.S. Foreign Policy* (1946), pamphlet.

9 Public Law 584, 79 Cong., August, 1946.

10 C. A. Siepmann, *Radio, Television, and Society* (New York, 1950), p. 299.

traveled all over the world and saw the lack of understanding of the United States and U.S. policies common among foreign peoples, that Congress began to recognize the need for organized, planned use of the psychological instrument. The Truman Doctrine had been enunciated and the Marshall Plan had been suggested. The congressional trips were an effort to test the usefulness of these two great postwar U.S. policies. Congressional conviction that these and other policies needed propaganda support to be fully effective resulted in passage of the Smith-Mundt Act[11] in January, 1948.

THE STATE DEPARTMENT AND SPO

This act gave the State Department's Office of International Information and Educational Exchange (OIE), successor to OIC, a legal status which the department's efforts in this field had not previously enjoyed. Gradually, the whole program increased, but it cannot be said the concept was fully understood or completely accepted. Both Mr. Benton and his successor, Edward Barrett, have spoken and written of the failure of old-line State Department and Foreign Service people to appreciate and work with this modern interloper in the quiet and dignified arena of diplomacy.[12] In 1951, as a senator, Mr. Benton was still strongly advocating the removal of the function from the Department.[13]

The line of reasoning of those who wanted the information program removed from the hands of the State Department was briefly as follows:

1. New ideas and flexibility in their use would be more likely in a new and separate organization. A new, different, and more imaginative type of person would be more attracted to the separate agency than to the State Department, which, justly or unjustly, suffers from the reputation of being conservative, staid, and stuffy.

2. The State Department is, and should be, the top foreign policy-making body in government. It should not be saddled

[11] Public Law 402, 80 Cong., January 22, 1948.

[12] E. W. Barrett, *Truth Is Our Weapon* (New York, 1953), p. 240.

[13] W. Benton, *How We Can Win the Peace by Improving the Voice of America* (February 19, 1951), pp. 8–9, privately printed pamphlet containing the Senator's speech to the Senate.

with *operational* activities which detract from the deliberative air of pure policy formulation. The proponents of separation pointed to the "wisdom" of Congress in establishing a separate agency for the conduct of the European Recovery Program (Marshall Plan). The psychological program is an even more complex operating problem than the ERP.

3. The State Department could and should continue policy control over psychological operations in the separate agency.

4. The State Department is not capable of, nor is it the proper place for, conducting those aspects of the psychological function which might be termed "unconventional propaganda."[14]

5. Though perhaps less openly argued, it was believed that a separate agency would not be subject to the slurs and financial cuts which Congress seems so often to be so free in using against the State Department.

The argument against separation ran along these lines:

1. Separation during the critical times of the Cold War would cause confusion and delay in prosecuting America's side of the struggle.

2. Since it is agreed by all that State must continue to have policy control, separation will make this process of control more difficult and will cause costly delay between determination of policy and its explanation abroad.

3. The interplay of ideas and personalities of the newer, more dynamic types attracted to the "information" field, and those of the older, more conservative, and traditional types within the diplomacy side of State has salutory effects on both.

4. The relationship of overseas personnel in the Foreign Service and in the new agency would be severely complicated to the possible embarrassment of each and of the United States.

There is validity in both arguments; and there is speciousness. There is no real reason why, if properly oriented and run, the program should not attract highly qualified personnel, no matter where its place in government. It is true that the State

[14] *Ibid.*, p. 10. By this Mr. Benton apparently meant everything less usual than "open-and-above-board information and education operations," but short of "dirty tricks" and "gray" and "black" propaganda which he felt belonged to the Central Intelligence Agency and the military.

Department is often maligned, but this is more because of its function than because of its people. The area of its activities lies outside the United States and therefore outside the ken of most Americans, including many in Congress. Much of its work is necessarily cloaked with some secrecy if it is to be effective. It has no "pork barrel" use to Congress, nor does it have any congressional constituents among its beneficiaries as in the case of departments such as Agriculture, Interior, and Commerce. But all these same things can be said of the "information" agency. Indeed there is the added disadvantage of its newness, lack of traditional acceptance, and the aura of something a little bit less than "nice" in playing about with men's minds. There has been, and will long continue to be, argument as to the wisdom of keeping the European Recovery Program out of State. Overseas the operational units had finally to be put under the chiefs of our diplomatic missions and, in some cases, the divisive forces between the two agencies became so marked that the chief of the diplomatic mission had to be named concurrently as chief of the Economic Cooperation Administration (ECA) mission.[15] While there is little disagreement on the foreign policy–making role of the State Department, it can logically be argued that it is also an *operational* agency since the diplomacy function is an active *operation* as distinct from the more passive representational and intelligence-collecting roles of the Foreign Service but even these, it can be said, contain elements of *operational* activity.

There is really only one valid question here. From what place can the psychological instrument be most usefully and fully employed? As has been discussed earlier, the problem hinges largely on the status position of the agency for SPO vis-à-vis other agencies in the Executive Branch, and of the agency's chief vis-à-vis the important individuals in Washington.[16] From that discussion, it would seem that a separate agency on the departmental level and a chief who is not only

[15] An early example was the case of Greece where the more usual problems were compounded by the Greek civil war. Mr. Henry Grady was named by President Truman to head the ECA mission while he was also U.S. Ambassador.

[16] Barrett, p. 246. "An important function requires important status."

a member of the Cabinet but a member of that "Cabinet within the Cabinet," the NSC, are essential. Policy control can be worked out, particularly if the SPO chief himself is a participant in devising the policies within the limits of which he and his agency must operate. The only other apparently feasible solution would be to create "junior Cabinet" posts within State as they exist within Defense.[17] There could then be a Secretary of Strategic Psychological Operations, a Secretary for International Economic and Technical Cooperation, etc., with direct responsibility to the Secretary of State but not so submerged in the official hierarchy of government as to be so little listened to on the Cabinet or NSC level as is the lot of the "Assistant Secretary of Something-or-other."

At this point some reader is certain to shudder at the thought of duplicating the "monstrosity" that is the present organizational structure of the Defense Department. It would certainly be safe to say that the organization has fallen short of what it was hoped it would accomplish. But the failure of the Defense Department to provide the United States with answers to the tactical and strategic requirements of the Nuclear Age cannot be entirely laid at the *organizational* door. It would seem evident that honest conceptual differences as well as jurisdictional disputes between departments of former Cabinet rank are partly to blame. A sometimes less than wise choice in political appointees, and a bewildering proliferation of unknowledgeable civilian superiors over career military men required by law to play out competingly dual roles must also bear some blame—perhaps a major portion of it. Organization per se cannot be expected to cure the ills resulting from confused and confusing missions and functions within it.

A Proposed Structure for SPO

The organization of an SPO agency that we believe could best fulfil the functions of the agency as outlined in the previous

[17] It is interesting that the Brookings Institution has suggested the establishment of a new "Department of Foreign Affairs" to supersede the State Department. This department would be divided into three "subdepartments" each headed by a Secretary of Cabinet rank. The three "subdepartments" would be for political affairs, for economic affairs, and for information and cultural ex-

chapter would have two characteristics which are different from anything that has been tried by the United States. First, it would be a cabinet level agency with its director a member of the NSC by statute. Second, in addition to the Washington and country organizations that are typical for any agency with foreign operations, it would have a series of regional headquarters. Much of the planning and intelligence functions would be decentralized and performed in these regional headquarters rather than in Washington.

In outlining the actual organization of the Strategic Psychological Operations agency no attempt will be made to discuss in detail any of the elements which are fairly normal to

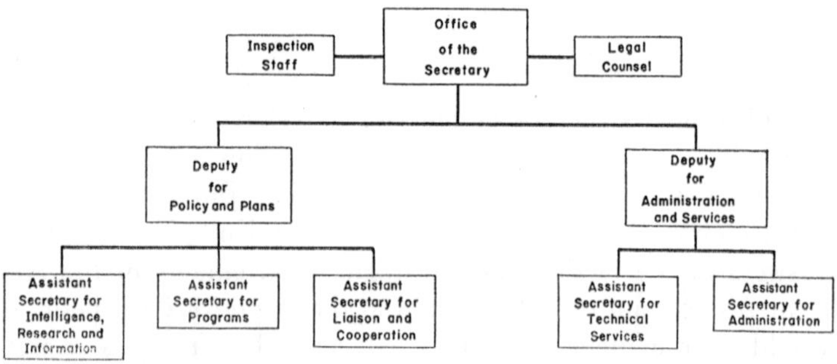

FIG. 3.—Proposed organization for a Department of Strategic Psychological Operations.

all agencies. The discussion of both the subdivisions and certain personnel will be confined, in the main, to those subdivisions which, because of the nature of their functions, would present problems which have not been fully understood or reflect points of view which have to date failed of general acceptance in Washington. In this same vein, the organization charts should be looked on by the reader not as large-scale, complete maps but as fairly accurate field sketches showing the relative positions of the major landmarks and principal points of reference.

The first and major breakdown of the proposed organization would look as shown in Figure 3.

---

change affairs. See U.S. Senate, 86 Cong., 2 sess., Committee on Foreign Relations, *The Formulation and Administration of United States Foreign Policy* (January 12, 1960). This study was prepared by the Brookings Institution.

The Office of the Secretary involves functions already mentioned in the discussion of the man and his place in the governmental structure. The Legal Counsel presents no unusual problems. The Inspection Staff, however, is something peculiar to the proposed organization. It would be composed of particularly qualified senior "career" men who are assigned to this staff on a rotational basis for a two- to four-year assignment. Dates and length of assignment should be so calculated as to give continuity to the operation. The inspectors, as are inspectors-general in the military, would be competent to investigate any and all aspects of the SPO agency's operations. They would be an extension of the Secretary's eyes and ears. They report directly to him and are responsible to no one on a lower echelon. It is this staff which would conduct periodic inspections of the operations of the various Washington activities as well as of those in the field. For reasons of effectiveness the men assigned to this staff must have demonstrated their own ability in various headquarters and field posts, and must be of unquestionable integrity and capable of objective judgment. An incompetent or biased inspector could easily and quickly destroy the respect in which this staff should be held, and might even wreck an operational activity through misunderstanding, lack of judgment, or lack of fairness. For the staff's own prestige and for the morale of those inspected, the inspectors must have had practical experience in the various echelons and in the field. It is this staff which continually reviews and analyzes the evaluation of effectiveness reports prepared by a branch of the Intelligence, Research and Information Office. This function was outlined in chapter iii. The Inspection Staff would use these reports as a basis for their inspections and, in consultation with appropriate policy and plans personnel, would make recommendations to the Secretary for improvement in a weak post or activity, or for emulation by others of a particularly effective program, activity, or post.

We shall treat with the Office of the Deputy Secretary for Administration and Services first because many of its functions would be more or less usual to any agency and can be briefly outlined. The Deputy would be responsible for two major subdivisions—Technical Operations and Administration. What is said of each of these major internal functions would be applicable to the Office of the Deputy and to his immediate staff. The Deputy himself should obviously have competence in administration, but more than that he should have experience in the administering of a large number of units and people geographically widely separated, preferably in a number of overseas countries. Since a large portion of his job would be the management of people working under difficult conditions, he should also be a man who is, through experience or imagination, able to understand the simple human problems which face Americans living in strange and remote areas of the world. Here, too, the need would be for a man experienced in mass media of communications, such as a world-wide news service or radio system. He should, if not himself a technician in one or more media, have enough technical background so that he fully understands what the technicians are talking about. He must be a man who is imbued with the spirit of service and who can impart that spirit to those working under him.

The Office of the Assistant Secretary for Technical Services, like that of Administration, would perform a service for the agency as a whole, and, like it, would report to the Deputy Secretary for Administration and Services. This office and the various divisions under it would be responsible for the physical operation of radio and television stations, for the operation of printing and duplicating plants and any other facilities concerned with the physical output of the product of the Strategic Psychological Operations agency. It would be in charge of the physical means of incoming data, for example, monitoring stations and other electronic devices, such as telegraph, telephone, and teletype wires owned or leased by the agency. It would run the internal communications services as well, including message centers, mailrooms, etc. Since relatively little of all this would be in Washington, its primary function in this area would be twofold: (1) to provide guidance and support to these technical operations in the field and (2) to maintain close liaison with private mass communication media and manufacturers of related equipment in order both to promote new ideas and to be abreast of the latest technical advances.

To carry out these functions would require three divisions. The Media Division would provide the guidance and support for radio, television, and motion picture facilities. It would also guide and augment the printing facilities in foreign posts. It should guide and support the preparation, transport, and display of exhibits. It might properly have a branch devoted to the development of facilities for special operations, such as propaganda balloons and wartime devices which would be inappropriate in peacetime. The Technical Services Division would provide the supervision and backing for the monitoring facilities. It would be the staff unit responsible for liaison with electronic research and manufacturing plants. The Division of Internal Communications would serve an essential, but self-explanatory, purpose.

The Office of the Assistant Secretary for Technical Services would need a highly skilled and experienced technical staff. These technicians must, moreover, be experienced in the facilities, radio systems, telegraph systems, printing plants, etc., of many foreign areas. They would often have to be improvisers and they should be men of sufficiently flexible minds that a failure or lack of the "usual" or "normal" facilities would not stop them for long. There must be inventors and technically imaginative people here too, not only to help grasp new devices and systems, but to suggest these to manufacturers and scientific researchers. There is room for a few "Buck Rogers" types in this side of SPO.

The internal administration of the SPO agency would not be particularly remarkable or unusual. It would face the same problems as those of other agencies having the bulk of their operations and people abroad. There are, however, a few caveats and some points which deserve mention. First and most important, this office must never be allowed to become the tail which wags the dog. This is a service function and not a directive one. All personnel must be made to realize this philosophy and to keep it in mind. Because it is a housekeeping function, a larger proportion of its people would be stationed in Washington. Since they are thus close to the sources of power, there would always be a tendency for them to claim for themselves, perhaps unconsciously, some of the "authority" of "Washington." Another tendency which should be avoided is the businessman's concern with profit-and-loss statements and accounting. Government is not a business; it does not work for a profit; "losses" do not exist in the business sense. There can be waste in government, to be sure, and it should be avoided, or at least minimized. But many governmental functions, and particularly those related to foreign affairs, are necessarily "risks" in the sense of "paying off." The risks are usually "calculated," but there are so many variables and so many unknowns that no agency and no government can hope for more than partial success in the sum of its foreign policy ventures. The factors of variables and unknowns are even greater in the use of the psychological instrument. We do know that it has potential; it has proved that in the past. We do not know just when and under what circumstances those potentials will be realized. The science of psychology is a new one and its application to masses is newer still. Cost accounting is a necessary evil of bigness and of modern life. Without question the Administration, the Congress, and the American people, as well as the agency itself, have a need and a right to know what any governmental operations cost. But there is a tendency in some quarters to let costs become a sort of negative guidepost and to let those who understand them best, the accountants, become some sort of director-guides. This tendency has been especially manifest in the military of late years, with the result that commanders no longer command, they now merely articulate and follow the directives of the super–cost accountants, the comptrollers. If the bookkeepers and the cashiers of the SPO agency are allowed to take over from the operations people, the agency had better never be established. For there is no way in which the effectiveness of this type of operation—the "profits" if you will—can be accurately measured. This is partly because some of the results may not manifest themselves for months or years afterward and partly because it is usually impossible to separate the SPO effects from those of other instruments which may have been applied individually or collectively to the same objective.

The Administration would of course have divisions dealing with personnel, with fiscal and budgetary matters, with property and purchasing, and with security. The fiscal and property functions speak for themselves. The security function would include both personnel security—investigations, screening, etc., which are essential to the substantive positions dealing in foreign affairs; and physical security matters—such as the safety of documents and facilities. It would also be charged with the security of communications, covering such things as couriers and code systems. Slightly more should perhaps be said about certain aspects of the Personnel Division. It would, naturally, deal with the processes of employment, promotion, training, assignments, and pay. More than that, however, because so many people of the agency would be abroad, many of them in remote and backward places, this division should have a branch devoted to something which might be called "Personnel Services." The primary function of this unit would be to serve the morale of the agency by making even the most remote employee feel there was "someone at home who cares" about him as an individual human being. All questions of emergency and medical leaves, dependents' travel problems, assistance in tax and legal matters could be referred by the individual to this unit for prompt and understanding attention. The matter of schooling for dependent children is a very serious one for a public servant who finds his are the only American children present in some foreign city. Special allowances or special arrangements, or both, will be necessary to keep this man content with, and free to work on, his job. Similarly there would be need, as the State Department long ago found to be true, for assistance in the simple matter of shipping goods and possessions from the United States to overseas stations, particularly if those stations are not on the normal freight routes of the world. All such things—many of them little things in themselves—can cause worry and harassment for the individual. Cumulatively they can mean the difference between high morale and low; between a job well done and one done indifferently or not at all. It is important that the Personnel Division should have in it personnel with some field experience. Without it they would never understand or fully tackle the problems which people in the field need solved for them if they are to concentrate on their jobs and apply their energies to their work rather than to the vexing problems of daily living in strange surroundings.

The Office of the Deputy Secretary for Policy and Plans

would be the real core of the agency in that it would constitute, other than the Secretary's office, the highest echelon for the translation of national policies into objectives and plans feasible within the capabilities and limitations of the psychological instrument of statecraft. It is here also that the development of new ideas would be born or would receive their final scrutiny before the Secretary proposed them to the NSC or other interested agencies of government. Furthermore this office would be on constant lookout for ways of increasing the capabilities and reducing the limitations of SPO. The office itself should be divided into a Policy Staff and a Plans Staff. Both must be manned with men and women of considerable intellect and broad vision. They need, in addition, many of the attributes and much of the background which are demanded of the "ideal" chief of the agency.

It would be false and entirely misleading to contend that all policy is made at the top. It is actually only the most important questions which are resolved there. The routine, daily work of all departments involves a myriad of minor policy decisions and policy interpretations. In the long run, these become policy. Such decisions and interpretations are made by "working level" people—the GS 13's to 15's and the lieutenant colonels and colonels who work in hierarchical anonymity but are nonetheless important in the overall policy framework. The SPO agency will have its share of these conscientious, hard-working, but unknown public servants who must work in close cooperation and coordination with their "opposite numbers" in State, Defense, and elsewhere. But this will become evident as the various internal functions and elements of the organization are developed.

The Policy Staff would keep in very close working relationship with the NSC staff, and with appropriate staff personnel at State, Defense, and other departments and agencies having functions in the international affairs area. This staff would set the policies for the regional headquarters and would provide guidance for them. In addition to the very real necessity for continuous initiative on its own part in the matter of contributing to the development of national policy, this staff should act as a two-way street for the rapid and easy flow of a traffic in ideas on policies and policy objectives between the NSC and the field operations. Like all "traffic," this needs some control, and this staff would exercise such control by keeping the field SPO people within national policy guidelines. But when convinced that these guidelines are no longer useful, because of either narrowness or misdirection, it would be this element which initiates a movement for changing them. The nature of this responsibility calls for mild conservatism from the performers of the function. Neither a radical nor a reactionary has a place here. Calmness, objectivity, analysis, and judgment are qualifications called for in the Policy Staff.

The Planning Staff, on the other hand, would need adventuresome thinkers and some "dreamers," as well as some of the more normal type of government career men and women. It should have a subsection whose sole purpose would be to think up ideas for new and different plans, strategies, and programs. The plans staff would provide a close working relationship with the OCB staff and the planners of other agencies. It would also work closely with the Office for Programs in the Washington headquarters and with the regional planners in the field. It is the plans staff which should develop war plans with the military departments and supervise the development and coordination of regional office war plans with those of the appropriate military theater commander.

A word should be said here about arrangements in times of general war. Under the American system, the military commander in a theater of operations has great discretionary power, far more, for instance, than his British counterpart. The wartime relationship of the SPO "country" representative with the military commander of a theater embracing that country would be parallel to the relationship between the diplomatic chief and the military commander. Briefly, if the government of the country in question is present on its own soil, in control of its own people, and friendly toward the United States, the diplomatic mission should continue to be responsive to its Washington headquarters, but should establish close liaison with the U.S. military theater headquarters. If the country is enemy-occupied, the diplomatic mission will cease to function and will probably be interned. If the country is "occupied" by U.S. armed forces because it was an unfriendly state, or because of the collapse of its own governmental control, the

U.S. diplomatic mission, if still present, should become a "political advisory mission" to the U.S. military government. In such a case the SPO agency representative would advise the military governor on all aspects of the psychological instrument while also maintaining his own agency lines of communication.

A separate branch should be provided for the development of special plans which usually have only a limited operational life but which may take months in preparation. This would include such diverse types of operation as those in connection with the 1948 Italian elections, the RFE (Radio Free Europe) operations, VETO and FOCUS, and Operation Mincemeat.[18]

It must be remembered, however, that both of these staffs, and particularly the Planning Staff, would not do "all the work." A great deal, and many times most, of the work of planning should be left to the field—both at the regional and at the country level. These are the people in closest touch with the audience country and its people. Those sometimes small events, or series of small events, which can change the mood of groups and, eventually, of a people and their government, are first noticed by men and women on the spot. Not only are they first noticed there, they are best understood there, and for that reason plans to profit by them or to counteract them can best and most quickly be formulated there. Changing situations (and the world's politics are ever-changing phenomena) and changing plans to meet these situations often cause a change in policy. Necessarily Washington must control policy, but it would be essential that the field, especially the regional offices, be given considerable latitude, within very broad policy limits, to amend or change plans previously approved in Washington. Minor policy changes must also be permitted for two reasons. Time is often of the essence in the application of proper psychological pressures to an existing or developing situation; and the authority to act on the spot and without seeking specific approval from a higher-up builds a sense of responsibility, an urge to initiative, and creates a challenge to the intellect which are all highly important to the effective application of timely psychological measures. Naturally, this is

18 Operation Mincemeat has been described earlier; the other operations will be dealt with in later chapters.

not to say that the country and regional posts would have a free hand to do as they please. Since what is done in one place always has repercussions elsewhere, there must be coordination in plans and policies; there must be a final, central control point. Only Washington, with its advantage of having the world-wide picture, and its knowledge of our world-wide objectives, can serve as that final authority. Obviously this arrangement would call for the most unfailing type of internal communications, between country and region, and between region and the Washington headquarters. Each higher level would have to be made immediately aware of new plans or new actions undertaken. The higher headquarters, also obviously, would retain veto power, but this power should be used sparingly. If the proper type of personnel are found for all the levels of the agency, there would normally be very little need to invoke the veto.

A warning is perhaps in order here. Any governmental structure must be arranged in echelons of importance and authority; the SPO agency is no exception. Nevertheless, this agency, maybe more than any other, must take care that authority is exercised in such a way that initiative and ideas have free and fertile soil in which to grow. No agency needs ideas more than one operating the psychological instrument of statecraft. Nothing can so effectively stifle the birth and blossoming of ideas in a government agency as the rigid and arbitrary application of hierarchical authority.

As the chart (Fig. 3) indicates, under the Deputy Secretary for Policy and Plans would come the offices of three Assistant Secretaries—for Intelligence, Research and Information; for Programs; and for Liaison and Cooperation. These will be treated in inverse order since there is a descending degree of difficulty in describing the organization and personnel involved in these functions. It is wrong to think of any subdivision of an organization as more "important" than another, since all are needed to make it possible for the organization to carry on its function. In this sense the horseshoe nail was as "important" to the battle as the tactical ability of the general, but it is certainly fair to recognize degrees of difficulty in recruitment for, and in the performance of, different functions. Looked at in this way the Office of the Assistant Secretary for Liaison and

Cooperation and its subdivisions would encounter the least difficulty of these three offices. This is not to say that the job would be an easy, or always a pleasant one. Nor is it to say that any man or woman picked at random could perform it equally well.

It would be the function of this office to establish and maintain close and harmonious contact with a wide variety of people and organizations in government and out. It would be the "public relations" element of the SPO agency, but this would mean considerably more than that term usually connotes. This would be so because of the great variety of "publics" with which "relations" are established and because of the existing, and previously described, misunderstanding and suspicion in those publics of the function of Strategic Psychological Operations. There is no product or service rendered which can be seen or consumed, or the effects of which can be measured or even directly felt. This office would have to convince its "publics" that there is an "apparent world" which may or may not be different from the "physical world," and that the former can and should be manipulated to the advantage of the United States as a whole even if no specific advantage accrues to the specific "publics" involved. Furthermore, it would have to be so convincing in this mission that it evoked voluntary cooperation from some of its "publics" to contribute their time, effort, and money to the national objective without hope of any direct return.

It is conceived that three major subdivisions would be necessary within this office. One, the Press and Public Relations Division, would perform entirely normal functions and need not be discussed. The second, the Liaison Division, would be responsible for liaison with both houses of Congress. This would entail establishing close, friendly relations with the key committeemen and other leaders in the ranks of both parties. This division would also be charged with the function of "cultural" liaison. Such liaison would mean close relationship with major university presidents and their social science faculties, and with various charitable foundations and church groups having an interest in world affairs. It would mean, further, a working relationship with liberal arts faculties, with major museums, art schools, and with the literary, musical, and dramatic worlds. Neither of these tasks would be easy; in one case because of resistance to the idea, and in the other because it is such a large and diffuse "public." The type of personnel needed for each liaison mission is quite different. Both would require intelligence, tact, a strong "sense of mission," and a certain ability to meet and talk with others. Aside from that, the proper people would probably be poles apart. Obviously the type needed for cultural liaison must be, themselves, highly cultured persons whose normal pace is geared to the relatively leisurely one of academia and the arts. They must have that blend of confidence and humility which so often marks the true intellectual. They should certainly almost totally lack the element of brashness which is essential to the liaison man with Congress. The latter must have more of the attributes usually associated with salesmanship. He should have persistence and a tough skin, for he will often find himself being given the "brush-off." A thorough knowledge of the workings of his agency and an ability to explain and, if you will, "sell," their importance would be needed. He must be punctual but have the patience to be kept waiting. He must be honestly a friendly person who loves his agency more than the "product," although he understands the latter. He would require a thorough understanding of the congressional system and an appreciation of the capabilities and limitations of it and of its individual members.

The third division would be that of Cooperation. Its function would be, as its name implies, to elicit the cooperation of both government and private organizations and individuals. A distinction should be made between "cooperation" by certain government agencies under certain circumstances, and coordination between SPO and the other instruments of foreign policy in the sense of integration for "total strategy." *Cooperation* would be called for, from both the agency and the individual, in the case of a Public Health Service officer who went abroad, for his and/or his agency's benefit, to study the conditions of Asiatic influenza in Burma. *Coordination* would be called for if the Department of Health, Education and Welfare sent a doctor or team of doctors to help the Burmese fight an epidemic of influenza. In the first case the SPO agency would have to learn of the trip from one of the interested parties. If the Cooperation Division had properly publicized itself, this would have been because HEW notified the division of the impending trip. In the second case the division, through its working relationship with other agencies, would have learned early along that Burma had requested such help from the United States. This division, in both cases, would act as the liaison agent between the SPO organization's planning and programing people and the HEW and the doctor or doctors concerned, in order that the psychological benefits to be derived, if any, could be fully exploited. Cooperation from many private organizations should be constantly sought. For example, the Ford Foundation in 1957 supported a study for the development of Delhi and a team of American "town-planners" went to India for over a year for the purpose of assisting in this work. The "psychological" benefit to be derived for the United States from such an operation is relatively small, but cumulatively many such undertakings have considerable impact. This potential will usually not be positively realized unless the SPO agency is aware of the activity and unless the people on the job understand the U.S. interests in the area to which they will go and the manner in which they may be able to forward those interests.

Book publishers and movie producers inherently have the means for furthering or injuring U.S. interests abroad. It is not advocating censorship to say that someone in government should make them aware of what good or harm they can do by their sales in foreign countries. Ways should be found for making U.S. books available to foreigners at a price, and in a currency, which they can afford. Only the surface of this potential has thus far been scratched; it needs deeper cultivation. If foreign audiences like our "westerns" and our "gang" films, and they seem to do so, we would be creating an Iron Curtain in reverse to deny their export. But ways should be found not only to export our better films at reasonable prices, but possibly to create greater demand for them among foreign audiences. The finding or developing of these "ways" would be a responsibility of the Cooperation Division. President Eisenhower's "Person-to-Person" program and similar activities of private groups and persons which are designed for greater international understanding, would also fall in the province of this division. The work of medical and church missions, too, is often fruitful psychologically. The work of such groups, however, can be most materially depreciated if it were thought by their beneficiaries that it was somehow mixed up with American "propaganda" or "psychological warfare." And, as a corollary, the psychological effect would also be negative. It is therefore necessary that great tact and subtlety be exercised in any such "cooperative" effort. In other words the cooperation must be "covert." This does not mean that it need be evil. Like propaganda itself, the effort at cooperation is "good" or "bad" depending on the means employed and the ends toward which it is directed. Another field in which this division would be, if not active, at least alert, and in which a maximum of discretion would be called for, is in the matter of relations with private organizations operating in the international propaganda field. Radio Free Europe, for instance, is an avowedly propagandistic effort directed at the Soviet satellite nations. To retain the advantages (including perhaps credibility in the minds of its audiences) of a non-governmental agency, it must avoid, and would resent, government interference. On the other hand, the ultimate objectives of that organization and of the United States are similar, if not identical. Furthermore the United States would (or should) be reluctant to have such a powerful psychological instrument straying too widely outside national policy limits. So long as it does an effective job and remains within those limits, it is to the government's advantage, as well as RFE's, to leave it strictly alone. In order to prevent a possibly wide divergence of policy objectives, some sort of discreet and covert liaison should be established. It would seem proper that the Cooperation Division undertake this.

It is difficult to pinpoint the type of personnel necessary to the Cooperation Division. Obviously they need tact and discretion. Equally clearly they should be willing to travel frequently. Since they seek and offer cooperation, they would themselves have to be cooperative people. They should be somewhat gregarious for they must meet, converse with, and win the confidence of, a great many very different kinds of people. They would need to be alert; their ears and minds should be constantly open to new ideas, new fields, and new means of cooperation. They must "know their way around"—because otherwise they would not be able to meet the proper people. Knowing the right person to see in government or in private organizations often means that one first must understand the highways and byways of organizational structure. The apparent, or constituted, hierarchy of any organization is often not the same as the hierarchy of real influence within the organization. Time can be saved and better results achieved by knowing who really wields influence at each level in every agency.

The Office of the Assistant Secretary for Programs might also be called the Office of the Assistant Secretary for Production, for it is within the limits of this function that almost all the other activities of the agency would culminate to achieve the "output." But "program" as used here is intended to convey the meaning also of "campaign," in the sense of a planned, integrated, and timed series of communications and actions designed to affect the physical and/or apparent worlds of the audiences. (See Part II for further explanation.) This does not mean that this process would all take place within this office. Quite to the contrary, the office in Washington should remain small, and most programs—or the product to be disseminated—should originate in the field. The Washington office would do some creative work where timeliness or proximity to the audience were not strong factors, but its principal function would be to furnish supervision, guidance, and support to the process in the various field posts. For that reason, the office should be divided along geographical rather than functional lines. What would be needed in the way of functional support would be obtained from other offices organized to furnish it. For example, additional radio wave lengths, changes in timing of broadcasts, or additional printing facilities would be obtained from the Office of Technical Services. Cooperation from the American Legion or from universities would be sought by the Programs people from the Office of Liaison and Cooperation. The production of a documentary film about some aspects of America or which, for some reason, could best be done here, would be supervised by this office, but would be a joint effort of the planners, the intelligence people, and the technical media staff. An added ingredient of such an undertaking would be the advice of the field post which requested it or of the posts in the countries in which it was planned to show the film.

There should, of course, be close daily contact between the programers and the policy staffs at each level. While the flow of policy communications between headquarters and field offices would be almost continuous and often heavy, the flow of communications between headquarters and field on program matters would be relatively light and often intermittent. It would often take the form of an exchange of ideas for new programs or comments on existing ones, with calls for, or offers of, assistance sometimes being included. This office would regularly review programs originating in the field. This would normally be after the event, particularly so in the area of radio scripts. Prior review might take place if a new program of particular importance or sensitivity were contemplated, or if Washington had doubts about the advisability of some planned program or part of a program. It would be this office, on its various levels and within its geographical subdivisions, which would work closely and continuously with "opposite numbers" in the Departments of State and Defense, and elsewhere, to achieve the final coordination and integration of the actual use of all four foreign affairs instruments.

The major subdivisions of the Office of Programs would parallel the number of regional offices. These divisions might properly be the West Europe, the field office of which might most profitably be in Paris;[19] the U.S.S.R. and East Europe with field headquarters in Munich;[20] the Far East with its field post in Manila; the Near East, Southeast Asia, and North Africa with field headquarters in Istanbul;[21] the Central and South Africa Division with a field headquarters in Brazzaville. The Western Hemisphere would be divided. A small unit, operating in Washington, would support the Canada country

19 Paris is a hub of European diplomacy and intelligence systems; NATO is there as is the U.S. military theater commander's headquarters. It is also well placed from the point of view of physical communication.

20 Munich has been a major Voice of America post. It is close to the Iron Curtain and is a refugee way station.

21 Istanbul has relatively good physical communication. Turkey is near the Arab world but, although Moslem, remains somewhat aloof from the internal struggles and conflicts.

post. The United States and Canada already have several joint activities in Washington and the relations are so close that no great problem can be foreseen here. In the case of Latin America, two divisions with parallel field regional posts would appear to be necessary to do an effective job. The Central America Division would have its regional post in Mexico City, and the South America Division's regional headquarters might be in Rio de Janeiro or Brasilia. It should be recognized that the selection of locations for all these regional offices is based on the world situation. Violent changes in the political situation of any of the areas involved, or major changes in the availability of physical facilities—radio, cable, airlanes, etc.—would cause a reconsideration.

There is contemplated also a Division of Special Programs which would concern itself with the development of programs the impact of which might not be relevant to any one geographic area, but would be useful or desirable in all or many of the areas of the world. Preparation of a documentary film on a purely American theme—*Public Education in America* or *The Tennessee Valley Authority*—are the type of thing in mind here and would be the responsibility of this division. Whether or not such films would be shown and under what circumstances and with what commentary, would be a decision of regional and country posts. Therefore the advice of the geographic divisions of this office would be sought by the Special Programs Division early in the planning of the film or any other project. Another example of the type of work which this division could undertake would be the USIA exhibits on Atoms for Peace which toured much of the world in 1955–57. The actual exhibits, however, would be constructed by the Exhibits Branch of the Media Division in the Office of the Assistant Secretary for Technical Services.

The personnel needed for such an office as this would represent in almost equal parts knowledge of their areas and knowledge of the psychological techniques of persuasion. These two qualifications should be fortified by a liberal quantity of imagination and a working acquaintance with the various media of communication. They must understand the planners and be imaginative enough to *want* to implement plans suggested, but they must also be aware of the likelihood of audience acceptance, taking into consideration the particular audience, the psychological validity of the "appeal," and the feasibility of means to convey it. Both the Intelligence and the Technical personnel would necessarily assist in such considerations.

THE SPECIAL PROBLEMS OF INTELLIGENCE, RESEARCH, AND INFORMATION

The Office of the Assistant Secretary for Intelligence, Research, and Information is perhaps the most complicated to organize because of the complexities of its functions. The underlying concepts have been dealt with at length in chapter iii. The organization implications of our concepts require brief discussion. The major portion of the functions assigned to this office would actually be performed in the field. The Washington intelligence function would be limited to a sort of "clearinghouse" activity and to supporting the policy, plans, and programs staffs in Washington. The clearinghouse function is essential in two regards: (1) it would enable all regional and country posts to be kept abreast of developments in areas and countries other than their own, and (2) it would enable the SPO agency to be a cooperating (and benefiting) member of the Washington "intelligence community." A further, and sometimes a most important, function of the Washington intelligence staff would be to establish and maintain an intelligence library which would be a final repository for intelligence documents.

The information function in Washington would be minor. Substantively it would serve only the policy, planning, programing, and intelligence staffs *in Washington*. Technically and administratively it would guide and support similar staffs in the field.[22] The "information" required by this kind of an operation normally falls into two categories, each of which requires a different sort of activity to produce it. The first required category is the information to support one's own propaganda output. The work involved is enormous for some kinds of propaganda operations. RFE's experience is that a tremen-

[22] The exception to this would be the Washington function of daily relay to all field posts of the "official" U.S. interpretation of events as evidenced by Presidential statements and State Department releases and statements. It is in this area that one of the reasons for close SPO-NSC relationships is to be found.

dous number of words—words from both sides of the Iron Curtain—is needed each day to be the grist for RFE's own word-producing mill.[23] The second required category is the less urgent (in the sense of timeliness) information to be derived from interviews with government and opinion leaders in all parts of the world, from such sources as public opinion polls, from magazine articles, and so forth.

Both categories are very closely related to output intelligence as discussed in chapter iii. "Output," particularly in propaganda, requires a volume of content—of words—day in and day out. *What* one says has to be largely based on *information.* The *way one says it,* the right words, so that the message does not jar the audience's ears as something patently "foreign"; the proper slant, so that the substance of the message has the desired "meaning" for the audience—these things are the result of output intelligence.

While *information* is necessary and usable in its original state, it is not *intelligence* until it has gone through the process of collation and evaluation. It is food for SPO in its initial form because it is "news." It is news that Nikita Khrushchev *said* he wants better relations with Yugoslavia, for instance. But one does not thereby have *intelligence* that the U.S.S.R. and Yugoslavia are moving closer together; there must be other evidence, perhaps a lot of it, before one can say this is actually taking place. *Information,* in addition to its own value in psychological operations, is also the raw material for the *intelligence* necessary to the effective use of all instruments of statecraft. Therefore the information staff of any SPO agency or field headquarters must be prepared to channel the flow of its intake directly to the policy and program staffs, and to the intelligence staff. There is obviously a time lag between information and intelligence. Sometimes policy and programs will find it inadvisable to use information which does not appear to jibe with previous intelligence on the same subject. In such a case they would call in intelligence people and demand

[23] Cf. R. T. Holt, *Radio Free Europe* (Minneapolis, 1958). RFE's *aural* monitoring (covering only a part of the world) produces 45,000 to 48,000 words a day; its *mechanical* monitoring produces 125,000 to 150,000 words a day. In addition to this source, words flow in by the hundreds of thousands a day, in the form of newspapers read and in the form of daily cables from its field offices on the periphery of the Soviet bloc.

a forced-draft sort of processing. It will often happen that the intelligence man will refuse the "forcing" because he will be unable to say "yes" or "no"; the best he will be able to do is to say, "in view of what evidence we have this information appears possibly (or probably) true (or false)." The policy and program people must then decide whether and how to use the item. This type of integrated action between the information, intelligence, and operating staffs will be a common—perhaps a daily—occurrence. It calls for people in each staff who understand the problems, capabilities, and limitations of the others.

This will not always be easy. Information personnel are newshawks; they must be alert, observant, and quick. They must sense where and when a news "break" is going to happen; they must anticipate and report events, but need not really think about the meaning of them. The intelligence people, on the other hand, should be considerably less news-conscious and far more conscious of the meaning behind the news. They are interpretive columnists rather than reporters. The city editor is the information man; the editorial page editor, the intelligence man. The analogy, like most, has some thin spots, but in the main it has some validity. It may help the reader to understand some of the needed differences within a closely operating group, doing different things but all serving the same end—the production of a newspaper—or the programs of Strategic Psychological Operations.

Before going further in this rather complicated function of SPO, it might be well to consult Figure 4 to show the proposed breakdown of this office in regard to the intelligence function.

The intelligence function, performed largely in the field and skeletonized in the Washington Intelligence Division, entails three basic activities. The library and the "clearinghouse" aspects have already been mentioned. The operation of the Audience Analysis Branch, broken down into sections paralleling the regional offices, requires some explanation. It is this unit which would undertake the relatively slower-paced and longer-range production of intelligence which might be called analysis of audiences and their situations—their situation as "audiences" within their own countries and the audience countries' situation in the overall world situation. It is this branch which

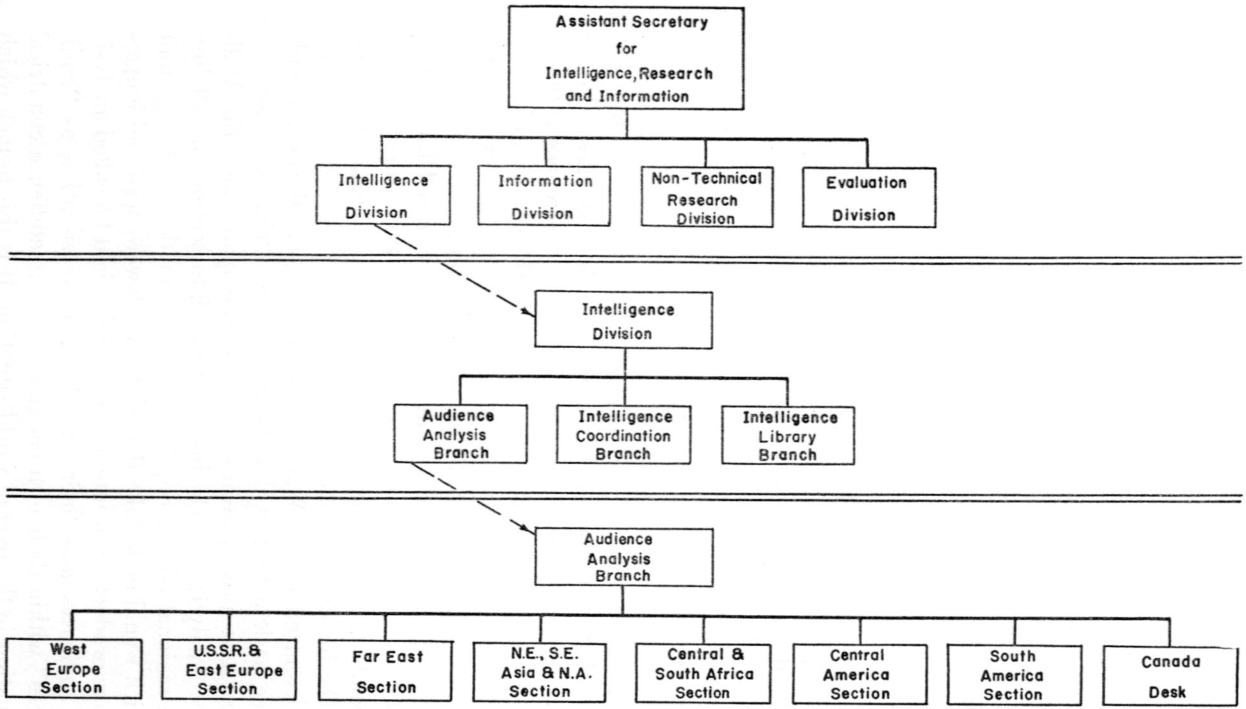

FIG. 4.—Proposed organization to deal with special problems of intelligence

would produce evidences of cruciality and susceptibility in audiences. What "audiences" in Italy, if appealed to, were most crucial to the re-election of Alcide de Gasperi in 1953? What were the susceptibilities of this audience?[24]

There are two modifications or clarifications which should be mentioned. First, in the field, particularly at the country posts, where the tie between SPO staff and embassy staff must be close, the continuing process of audience and situation analysis is somewhat different. It is a prime function of embassies to keep abreast of "the situation." Therefore the SPO man would gain much of his intelligence, in already analyzed form, from the political officers at the embassy. He, in turn, should be able, through his own analytical processes, to designate for them the "audiences" within the country which are both crucial and susceptible to U.S. influences. The second function is an enlargement on the Italian example used above. It has to do with what may tend toward duplication in an intelligence function, but which nevertheless has distinct aspects that, despite some overlap, should be kept separate. The Information Division (and its field counterparts) has an *intelligence* function of its own as distinct from *information.* The example used above related to an event which was foreseen and for which considerable prior planning could have been (and perhaps was) done. But the information function is one primarily of feeding the propaganda activities of SPO. As such it must be geared to speed and flexibility. There is frequent need, as world events unfold, for what is almost intelligence, but what may not quite qualify for that term, because of the thinness of its base, or because it is so current it has not been, and cannot be, tested against time. The "audience" may have to be changed suddenly because the regime has changed, because a coup d'état has taken place, because some momentous economic change has occurred. Or the theme of output may have to be changed because of some change of heart in the

24 While this example is picked because of its actuality in American "information service" history, it should be remembered that the real work involved in such analyses in the proposed SPO agency would be done at the field posts —in this case, largely at the country post in Rome, backed and supported by the regional office in Paris.

audience.[25] In the daily deadline atmosphere of propaganda output, it is impossible to wait for the orderly and considered processes of the Intelligence Division (and its field counterparts) to come to carefully evaluated conclusions. A subfunction of the information function—as it is always *au courant* with day-to-day developments—must be to designate the new most crucial and most susceptible audience. It would be this element's function also to furnish the propaganda producers the names of newly emerging important people and the newest slang and idioms to appear in their news gleanings. These people must hear, understand, and pass on to the producers such terms as "Teddy Boy" and "cult of the individual" long before such items would appear in the normal intelligence reports. Nothing so clearly raised the warning flag "propaganda" to American soldiers during World War II, as German, and particularly Japanese, misuse of American slang and idiom. Even though these broadcasts were from non-concealed sources, our men listened to them for music and for entertainment. The effectiveness of them as psychological warfare was nearly nil. One principal reason for their conspicuous lack of success was that almost without fail every enemy broadcast managed to destroy its grip on its audience by the inclusion of some word or phrase which jarred the American listener back to reality and which usually made him laugh at "them Krauts" or "them Nips."[26]

Objective evaluation can only be conducted by persons other than those whose work is being evaluated. This means that evaluation in the field should be the function of the regional unit and not of the country post. To be thorough, however, evaluators would have to be pretty steadily in the audience country. This could be accomplished by "attaching" an evalu-

[25] Take again the example of the 1953 Italian elections. Suppose in the midst of the campaign some prominent labor leader in Turin appears to have swayed away from the Communist party a considerable audience in a normally CP stronghold. *Information*, not *intelligence*, must come up *the same day* with the answers to "What did he say?" "How did what part of his audience react?" Intelligence, possibly a day later, may answer the question, "Why was it effective on this audience at this particular moment?"

[26] One of the authors listened, and watched men listening, to hundreds of enemy propaganda broadcasts by both Tokyo Rose and Axis Sally. Despite sexy voices, popular music, and sometimes interesting information, the total effect was always as indicated above.

ator from a regional headquarters to an *embassy*, and not to the SPO country post he is evaluating. For audience countries which lie behind the Iron Curtain, there may be difficulties in thus attaching additional personnel to an embassy staff. And even if this were feasible, the restrictions imposed on travel and contacts within the countries would probably nullify the effort to evaluate. In these cases, the evaluators would be almost entirely restricted to those sources which are available. These include propaganda and counterpropaganda of the audience country itself, for these are frequently reflections of the impact, or lack of it, of our own psychological operations. They also include the interrogation of refugees and interviews with non-refugees, such as conducted by American newspapermen who have recently come from the audience country. Actions themselves sometimes give indications of effectiveness. The organization of Communist youth groups into leaflet-gatherers, with appropriate rewards, for example, is an indication of the effectiveness of the balloon-carried leaflets which the Free Europe Press sent into Czechoslovakia. Fan mail is indicative and, despite Iron Curtain difficulties, continues to be received by RFE and the Voice of America.

There would be a constant temptation to use evaluation reports as justification for one's operations. This temptation would be particularly strong in the face of congressional antipathy, especially at appropriations hearings. Properly used and honestly "hedged," such reports would appear to have limited use in such circumstances. But as has been said repeatedly before, the cause-and-effect relationship of SPO to "audience" behavior is extremely difficult to measure and impossible to prove. Too extreme claims by ardent advocates of "psychological warfare" have harmed rather than helped the concept in the past in the eyes of those whom it was desired to win over, both in Congress and among the military.

The Non-Technical Research Division, with no counterpart in the field, would nevertheless be a small division since it is not contemplated that much actual researching would be done regularly within the agency. The primary mission of this division would be threefold: (1) to keep abreast of research developments here and abroad in such fields as social psychology. communications (in the sense of conveying meaning to include such things as rumor), and the development of mass persuasion methods; (2) to prepare reports on the results of this research in non-scientific language, readily understandable by the layman operating personnel of the SPO agency; and (3) to stimulate, and in some cases to contract for and supervise, research on specific questions raised by, or apparently troubling, the operators. A typical question in the social psychology field which might be asked and the answer to which is obviously important, would be, "Why does an initial improvement in living conditions among some poverty-stricken and backward peoples turn them toward rather than away from communism?" This was one of the earliest results of the improvements made in the living conditions of the very poor in southern Italy. An answer to such a question would be of great importance because it would indicate either different initial methods or avenues of follow-up which would counter this apparently natural phenomenon. The personnel of this division should be highly qualified social scientists. They should understand scientific "jargon," but they must also understand the SPO operator's English. They should know enough about SPO to understand its needs and they should know enough about the state of social science to understand its capabilities and limitations. They must know, or know of, the leaders in these areas of research. Through study and personal contact, they should remain aware of new avenues opening up in the direction and scope of social science research. If they are not sufficiently scientific themselves, they will be largely useless and if they are not also fairly competent in the SPO field, they will be "used" by scientists as a source of funds with little promise of return to the SPO agency.

ORGANIZATION IN THE FIELD

Figure 5 gives the organizational structure of a typical regional office of the Strategic Psychological Operations agency overseas. Individual country posts will vary widely depending on the size, importance, and situation of each country. No description of a country post will be given, but it should be borne in mind that, no matter how compressed—perhaps even to a handful of personnel—the *functions* in the country posts will necessarily be parallel to those at the regional offices.

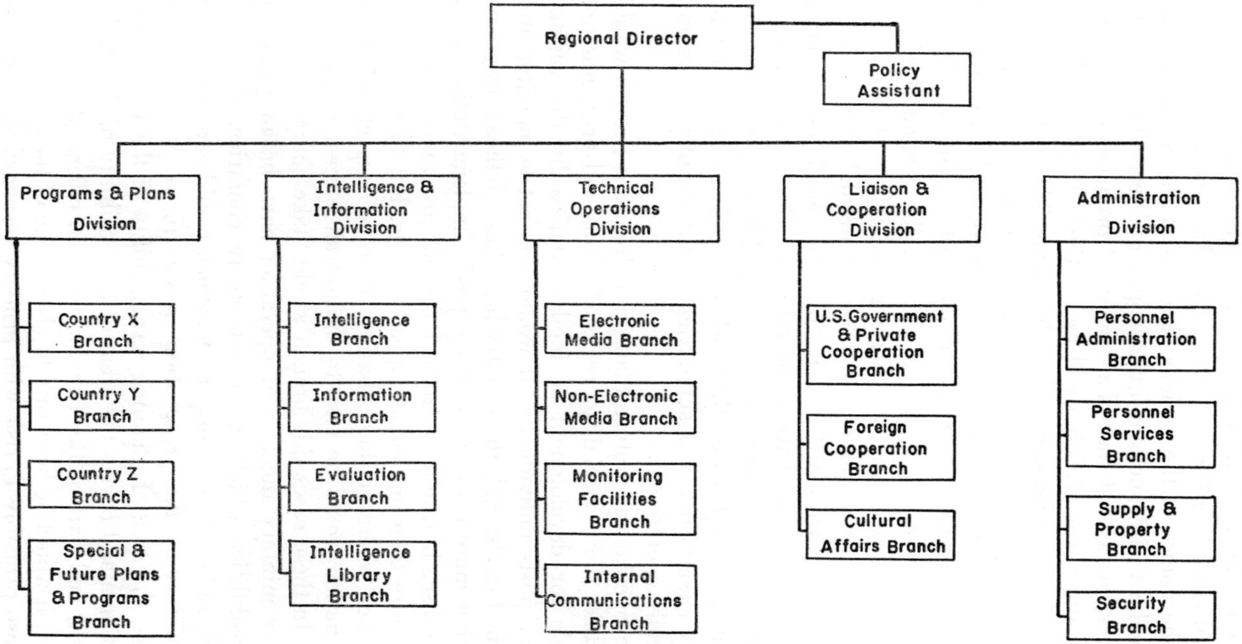

FIG. 5.—Proposed organization for major field operations

Little needs to be said about the functioning of the subdivisions of the regional office. In the light of what has been covered regarding the functions of the Washington headquarters, the names of the various elements at the regional office will make their functions quite apparent. Several comments, however, are called for. The "country" branches of the Programs and Plans Division will be relatively large when those countries are in the so-called denied areas. This is because the major operations aimed at those audiences would have to take place elsewhere than at the country posts. This would mean that all these branches at the Munich Regional Office and some of them (China, North Korea, etc.) at the Manila Regional Office would be large. At other regional offices they may be small and, in some cases, even "regionalized." For example, at the Paris Regional Office it might be found both efficient and economical to have a Scandinavia Branch rather than separate Norway, Sweden, Denmark, Finland, and Iceland branches. In such cases the work of the branches would be largely supervisory and supporting of the country posts and not really operational.

The Special and Future Plans and Programs Branch would undertake those special, or regional, plans and programs not suitable for development or implementation by the various country posts individually. It would also act as monitor and coordinator, through the country branches, of those programs requiring more than one country post's implementation. In a loose sense, it would be normal that all programs would require more than one country post's efforts. But there are special cases where timing, phasing, or other aspects of close coordination would be vital to the overall success of the program. In these cases this branch would "take charge."

Since military theaters of operation have broad planning responsibilities which encompass many countries, it is this branch, rather than country posts, which would cooperate with the military in the preparation of war plans. Where, as in Europe, the area of planning responsibility of the U.S. European Command covers the areas of responsibility of the Paris, Munich, and part of the Istanbul regional offices of the SPO agency, it would be necessary to charge one of these offices (in this case probably Paris) with primary responsibility and assign supporting roles to the others. It is evident from the foregoing that this branch (and its division) would have to have two points of supervision in Washington; the Plans Staff for war planning and similar matters dealing with plans for the long-range future and the Office for Programs, both for "regional" and "special" programs.

Another regional function requiring special comment appears to be in the realm of the Information Branch. In addition to the continuous flow of information which would be normal from country posts to the regional office, the latter would require special sources of information. This would be true almost everywhere, but especially true in connection with countries of the Soviet-Chinese bloc. It would be necessary to establish small offices in suitable places, as Radio Free Europe has done, for the collection of additional information.[27] The supervision of these field collecting-posts would be the responsibility of the Information Branch.

The last regional function on which a special word is appropriate is in regard to cooperation. The scheduling of, and arrangements for, the arrival of a U.S. fleet unit in a series of foreign ports would be the responsibility of Liaison and Cooperation Division. This division would also act as ultimate agent between a town in France, for example, which wanted a U.S. military band or formation for a commemorative ceremony, and the U.S. military in Germany who were capable of furnishing what was needed. This division, and its counterpart function at country posts, would be the one to promote means of greater understanding between our military forces stationed abroad and the sometimes reluctant hosts in the receiving country.

The principle that there can be but one boss at any one foreign post is sound and should be maintained. Of course if SPO is a State Department function, the question of who is in charge of the psychological operation at a foreign capital does not arise. If SPO is to be separate from State in Washington, there may be some question, although it is believed it is so easily answered it should not have arisen. The chief of the diplomatic mission—the ambassador or minister—must be the

[27] And which RFE now uses; cf., for example, Holt, chap. vi.

senior U.S. representative. The senior man from the Strategic Psychological Operations agency works with, but under, the diplomatic chief, just as do the various attachés. Like the latter, the SPO agency man has his own separate lines of communication and administrative control. Head-on conflicts, which sometimes do occur between men both honestly attempting to perform their jobs under these conditions, would be referred by each to his respective Washington headquarters for mutual settlement there or further reference to the Executive Office or the President himself for decision. This is now the system in effect in regard to the attachés of all military services and other agencies having representation at diplomatic posts. It would continue the present system under which the USIA representative abroad is the embassy's Public Affairs Officer (PAO), and therefore a member of the ambassador's staff. It has been proved workable and cases of disagreement requiring Washington's intervention are relatively rare.

• PART II

INTRODUCTION

In this section we will present three case studies of the use of the psychological instrument in the fairly recent past. Two of them deal with different phases of United States–Italy relations. The first is set in a wartime situation; the second in peacetime, or as nearly a "peacetime" situation as the Cold War seems to permit. It is believed useful to present them as two separate studies, since in one a single man with almost complete authority for the area, the commander of the U.S. Mediterranean Theater of Operations, had virtual control over the uses not only of the psychological instrument of statecraft but of the other three instruments as well. In the second, the circumstances were more normal in that once again both the Congress and the Executive Branch had resumed their authority over various aspects of international relations.

It is hoped that these case studies will be particularly useful to the person who has had no experience with the use of the psychological instrument. They illustrate such things as the large number of different media employed, the use of various techniques, and the problems of integrating the psychological instrument of statecraft with the other instruments in a major campaign. These are matters that we have not dealt with directly in Part I because they were somewhat outside the area of major concern and can be treated much more clearly in a case study.

The third case study we present as an example of the use of the psychological instrument by a private, non-governmental agency, Radio Free Europe. In some ways it is not entirely appropriate in this volume, because it represents a psychological operation undertaken by an agency that is incapable of using or of even materially influencing the use of the instruments of diplomacy, economics, or military force. But it may be the most useful of the cases because it alone represents a propaganda campaign developed and executed by a single agency. This is the only one of the cases in which all of the policy directives and evaluation reports were available to us for study. All *governmental* use of the psychological instrument is necessarily shrouded in secrecy. This has made the task of compiling the Italian cases more difficult. Furthermore, the actual records, even if "declassified," are often scattered and fragmentary. In the case study of Radio Free Europe's operations VETO and FOCUS, the directives were available and it is possible to illustrate how RFE analyzed the situation and developed propaganda policy to deal with it.

• CHAPTER V

American Psychological Operations in Italy 1943-45

Never is it easier to integrate all the instruments of statecraft, insofar as their impact on a foreign population is concerned, than it is in time of a major war under the American system of "military theater command." Under our system the military theater commander, while of course still subject to orders from Washington, translates those usually broad policy directives into action on the military, political, economic, and psychological fronts. He is generally secure in the knowledge that his Commander in Chief and his nation (through Congress) are united in a determination to "win the war," and in their acceptance that "the man on the spot and responsible for waging the battle knows best how to conduct it." Because this is so, it is particularly unfortunate that, as a people, we are still so little aware of the rest of the world's fears and aspirations. If we were, the broad directives to our theater commanders would be less broad and more specific, and our theater commanders would more effectively utilize their almost unique ability to accomplish real integration of the four instruments at their disposal.

Wartime Italy and the Exercise of American Statecraft

The invasion and liberation of Italy provides perhaps the best example for testing these contentions. This case study of the use of the psychological instrument in Italy in the years 1943–45 will, we hope, show how the instruments were all being used concurrently. It will also show, we believe, how the effectiveness of the psychological instrument (here limited by circumstances to "propaganda" and "psychological warfare") was blunted by the misuse, through ignorance and shortsightedness, of the political instrument.

This period was one of great mental and emotional strain for the "audience," since it included a slow-moving and devastating war which saw their country emerge from native dictatorship to undergo a harsh German occupation, which in turn gave way to a lenient Anglo-American occupation and finally complete liberation. The status of the country changed from that of active Axis partner, to a divided, defeated nation, to an active cobelligerent on the Allied side.

The "invasion period" will be both easier and more difficult to study than the postwar period treated in chapter vi. It will be somewhat easier because both propaganda and actions having a psychological effect were largely limited to the military insofar as contact with the Italians was concerned. It will also be somewhat easier because certain official records[1] were made available to us. It will be more difficult because there are almost no Italian records for that period available. They probably do not exist because of the disruption caused by war. It will also be more difficult because both propaganda and activity directed toward Italians and carried on in Italy were Allied rather than American. The sorting out of that which was purely American may sometimes have to be more inductive than deductive.

THE MACHINERY FOR PSYCHOLOGICAL OPERATIONS IN THE MEDITERRANEAN THEATER

The supreme command for the Mediterranean area was Allied Force Headquarters (AFHQ). While General Eisenhower commanded AFHQ he was not only overall Allied commander, but

1 AFHQ-PWB, *Summary of Psychological Reactions,* published daily and containing a summary of Italian propaganda beamed to the Allies, to Italians, and to the Axis. AFHQ-PWB, *Directive,* published weekly and containing the specific appeals and themes to be used in propaganda beamed to the Italian people—hereafter referred to as PWB-*Directive* (with date of start of week).

the senior U.S. military man in the theater. When he left to go to England late in 1943, General Jacob Devers became the commander of the U.S. Mediterranean Theater of Operations (MTO), while the Allied command went to a Britisher, General Sir Henry (Jumbo) Wilson. Under both Eisenhower and Wilson, however, AFHQ and its Psychological Warfare Branch (PWB) remained truly Allied enterprises. PWB contained American military and civilian (OWI) personnel and British military and civilian personnel, all working as a unit.

With PWB, as in most undertakings among allies, it is probable that the product at times failed completely to satisfy one or both of the partners. It can be assumed, however, that the propaganda emanating from AFHQ did not run contrary to the national policy of either nation insofar as such "policy" actually existed. For this reason, it is safe to say that it was "American" in its overall aspect. In fact, both allies from time to time did disagree with the direction propaganda was taking, but these differences were naturally worked out behind the scenes and were not made apparent to the Italians. It should be remembered that PWB was the source not only of psychological operations directed toward Italy but also of that intended for North Africa, southern France, the Balkans, and the Near East, as well as those areas of Central Europe which its radio and leaflet messages could reach. During the period in question, the major effort was, of course, that devoted to what can be called strictly psychological warfare—the attempt to weaken the fighting spirit of enemy troops facing the Allies. The most important civilian propaganda was that directed at the Italians on whose land those enemy troops were making a stubborn defense.

The makeup of PWB indicates the scope of its activities. To furnish itself with current information needed about its audience, PWB had a main subdivision known as "D" Section. "D" Section's major duties were to augment the normal sources of intelligence and information by gathering from local inhabitants data about "political personalities, currents of opinion, hostile or dangerous elements, economic conditions, local grievances, and so on."[2] The propaganda unit was called "E"

[2] J. A. Pollard, *Preliminary History of Foreign Operations, Office of War Information, to the End of the Italian Campaign*, p. 44. Typescript loaned to

Section; its activities were divided among various divisions whose names indicate their functions: the News Division, the Radio Division, the Films Division, the Publications Division. The actual operations and products of PWB will be detailed as the story of the psychological operations effort in this period is developed. It must be remembered that the account will be limited to propaganda and related activities directed toward the Italian *nation* and the tremendous effort of PWB in regard to enemy troops in the field will be largely ignored.

PWB moved to Naples only in August of 1944, but a branch was moved to Tunis shortly after its capture in order to be that much closer to Sicily and Italy. These were in a sense "rear" headquarters because PWB teams accompanied the invasion forces to Sicily and Italy and worked closely with both field commands on psychological warfare and with Allied Military Government (AMG)[3] on propaganda addressed to the civilian population. It is the latter type which particularly concerns us. Some mention should be made of AMG problems and policies in order to understand the importance to it of close cooperation with propaganda agencies. AMG itself, inasmuch as its American officers were the closest contact with America for many Italian citizens, was an aspect of psychological operations. It "played an important part in American foreign relations."[4]

MILITARY GOVERNMENT

To fulfil the requirements of international law as well as to protect itself from hostile activity in its rear and thereby to assure its lines of communication, an invading army must institute control of the civil population of the invaded territory.

R. W. van de Velde by Professor Pollard, who was for a time official historian of OWI. Many of the details as to numbers of radio listeners, numbers of newspapers published, dates of PWB actions, etc., which are found in this chapter come from Professor Pollard's work. They are too numerous to mention individually in notes. Our gratitude to him is enduring. He will be further cited only in those places where the reader might be misled into thinking a preceding or following citation is the source of the material given.

3 For a time called AMGOT (the OT standing for Occupied Territories). It will be referred to in this study as AMG for clarity and brevity.

4 H. Holborn, *American Military Government* (Washington, 1947), p. ix.

This is a complicated task, especially when the regime which is being supplanted is of a form to be obliterated and when a normal state of poverty has been tremendously accentuated by the rigors and material destruction of war. Both these added burdens were true of Italy. Fear, confusion, lack of communication, lack of food and distribution services, lack of health facilities, damaged or destroyed water supply and sewage systems, lack of fuel and electricity, masses of refugees, orphans and destitute persons—all added to the purely governmental problems facing AMG as it attempted to stabilize the populations immediately in the rear of the fighting troops. Getting at least the bare facts of the situation and making known to the bewildered population the new regulations and measures to be undertaken fell to the lot of those PWB units working with AMG.

Fairly early in the war the Army realized that it would be faced with the problem of military government. The United States, essentially an isolationist nation prior to the war, had no trained corps of foreign administrators to draw on, as did Britain, for example, with its empire. At Charlottesville, Virginia, a School of Military Government was established. Other universities were later enlisted to help round out the program. The courses were brief and largely technical; most of the students (officers) were lacking in any appropriate background. Under these circumstances it is remarkable that their record in the field was as good as it was. It has often been said that almost total lack of awareness of the sociopolitical problems of the occupied areas was normal among AMG officers. That seems to be the fault of basic American traits and institutions rather than of the Army.

SPECIAL PROBLEMS IN ITALY

AMG was the Army's, and therefore the Allied governments', instrument for political and economic control of areas when combat troops had taken them. In a strictly enemy country it would probably have remained in charge until a political settlement had taken place at the end of the war. In Italy the situation was different. The Allies claimed to be fighting fascism and not the Italian people. Fascism had fallen and steps had been undertaken by the new Italian government to remove traces of the Mussolini regime before the fall of Sicily and before invasion of the mainland. Moreover that invasion was linked in timing with the surrender by Italy to the Allies.[5] When Italy declared war on Germany and became a cobelligerent of the Allies in October, 1943, it became even more obvious that the rear areas, removed from the fighting, should be treated in some way different from those generally applied to conquered peoples.[6] At the same time it seemed advisable, in view of the long reign of fascism and the apparent divisions among the Italians themselves, that some form of Allied control be retained for a time. In November, 1943, General Eisenhower created the Allied Control Commission (ACC), with himself as ex officio president. A Chief Commissioner was the working head of ACC and in January, 1944, this official was made AFHQ's Chief of Civil Affairs, thus assuring coordination between ACC and AMG.

The difference between the two organizations was a real one. AMG worked under the jurisdiction of troop unit commanders as far down as a division. ACC's authority began at the rear boundary of the Fifth and Eighth Armies' areas—in other words, in those areas where Italian civil authority had come into limited being. Its functions were to "enforce the terms of the surrender [never too clearly stated], to organize Italy for active participation in the war against the Nazis, and to assist the growth of democratic life."[7] Without minimizing the problems facing ACC, it can be said that AMG's problems were more pressing, more varied, and often more difficult. Nevertheless the two together constituted Italy's first taste of the policies (or lack of them) of the Western Great Powers. The role of AMG and ACC, often misunderstood by the Allied commanders, by the Italians, and even by the civil affairs officers themselves, may have been more lastingly effective for good or ill than the task of fighting the war.[8] It will be essential to this study to examine that role, as well as the propa-

[5] D. D. Eisenhower, *Crusade in Europe* (Permabook ed.; New York, 1948), pp. 210–13.

[6] Cf. PWB *Directive*, October 18, 1943.

[7] Holborn, p. 20.

[8] Cf. E. Sevareid, *Not So Wild a Dream* (New York, 1946), pp. 383, 423.

ganda of PWB which was in many ways the mouthpiece of AMG-ACC.

SPECIAL PROBLEMS FOR AMERICANS

There seems to have been, especially in the American forces, an inherent discordant note between the military leaders and civil affairs officers. The average American lacks knowledge of foreign peoples and of political ideologies and economic systems other than his own. The American characteristic desire to "get the job done and finish with it" was a trait to be found in most U.S. commanders. On the other hand, the very nature of civil affairs work makes it largely political and economic. In Italy, it could not avoid clashing ideologies any more than it could be made to fall into neat and ordered compartments like the more usual military duties. Nor are the Italians a people to be normally rushed into things by American exhortations to "get on with the job." Military manuals are necessarily cut-and-dried texts designed as fundamental rules; they cannot foresee or guide in all contingencies. Directives could be and were promulgated that such Fascist laws as those dealing with race, religious discrimination, and political privilege were to be abolished. But the day-to-day problems in individual situations had to be faced and solved on a basis of common sense within a very broad area of generalized policy such as to "restore law and order." What laws and how to appoint those to maintain order were usually left to the discretion of the civil affairs officer on the ground. These are very real and pressing problems to a lone American with little experience in government or with foreigners, and yet this was only part of the task facing every civil affairs officer entering a liberated town to "take over" for his own government. It is interesting to note that three major U.S. commanders who were at one time concerned with the Italian campaign and who wrote books on the subject showed little awareness of the political aspects of that campaign as it would affect postwar Italy.

The Psychological Instrument

PROPAGANDA POLICIES APPLICABLE TO ITALY

Elmer Davis, head of OWI, outlined the basic propaganda policies of the United States in regard to enemy, occupied, and liberated countries.[9] All three applied at times to Italy. He said "the objective . . . in enemy countries is to convince them that their defeat is certain; and in so far as national policy permits, to convince them that, in the words of the President, the conditions of peace will permit them to have an honorable and useful existence after the war and after they have purged their militaristic [or fascistic] elements." In occupied countries, said Davis, "we do our best to keep up the spirit of the people and urge them to resistance; but we keep telling them not to rise up too soon and urge them to wait for instructions from their governments in exile, who will tell them when the day comes to offer active resistance." Whereas in liberated countries "it is largely a question of getting people to cooperate with the Army of Liberation and go along with our side; and of reconciling them to the fact that liberation is not a matter of getting everything they hoped for all at once."

In the final days of the Tunisian campaign and prior to the invasion of Sicily, the emphasis in propaganda to Italy was on the strength of the Allies, the inevitability of defeat for Italy. The Allied armies were pictured as an irresistible liberating force.[10] These two themes had been in effect ever since the landings in North Africa had been achieved. The luster of "liberation from fascism" theme had been more than somewhat tarnished by the Darlan deal some months earlier, of which Edward Barrett said, "American cooperation with Darlan in North Africa caused all but *rigor mortis* in the psychological offensive."[11] This act of expediency will long be discussed. It is not our purpose here to reopen the argument, but it is necessary to point out that the Italians were not ignorant of the facts and that the knowledge probably tended to confuse

[9] U.S. Congress (House) *Hearings on National War Agencies Appropriations Bill for 1945* (Washington, April, 1944), p. 17. This report will hereafter be referred to as House, *Hearings 1945*. Note in this "policy" the lack of any mention of the postwar objectives; everything is implied to end with the end of the war.

[10] U.S. Congress (House), *Hearings on the First Supplemental National Defense Appropriations Bill for 1944* (Washington, 1943), p. 1291. This report will hereafter be referred to as House, *Hearings 1944*.

[11] E. W. Barrett, *Truth Is Our Weapon* (New York, 1953), p. 43.

them, as the facts did not match statements such as that made later by the Secretary of State and duly publicized that "there cannot be any compromise with Fascism—whether in Italy or any other country."[12] It was this sort of discord between policy and action which led observers to note that the Italians were at a loss to decide whether Americans were liberators or conquerors.[13] In this regard General Eisenhower made a telling point on the dangerous repercussions of opportunism in military strategy, "so often applicable in tactics," but strangely ignored the parallel in political strategy.[14] The Italians were accustomed to expediency in their own government, in Hitler's Germany, and in Stalin's U.S.S.R., but it was a new and depressing thing to find the United States of noble words like those of the Four Freedoms and the Atlantic Charter also taking such measures.[15]

There were other problems to overcome. Some of the British military wanted to broadcast a report of an armistice between Italy and the Allies concurrently with the news of Sicily's invasion. This is an example of the type of well-meaning suggestion the propagandist is apt to encounter from those who see a possible immediate advantage but fail to weigh the longer-term results. This would have been entirely false propaganda which would almost at once have been revealed as such and would have undermined the carefully nurtured reputation for truth which Allied information agencies had built up.[16] Fortunately the professional propagandists prevailed and the idea was dropped. The invasion was announced in a brief and truthful statement saying that the invasion had commenced and that the Italians could have peace if they wanted it. This announcement was made over the radio and through literally millions of leaflets. An indication of the usefulness of truth in

[12] C. Hull, Statement on U.S. Foreign Policy, April 9, 1944, printed in OWI's *Victory Bulletin* (April 24, 1944), pp. 3–4.

[13] Cf. Sevareid, p. 365.

[14] *Op. cit.*, p. 186.

[15] Cf. G. de Santillana, "Italy Listening," *Atlantic*, CLXXI (March, 1943), 66–72.

[16] Cf. Bruce Lockhart, *Comes the Reckoning* (London, 1947), pp. 249–50; and H. C. Butcher, *My Three Years with Eisenhower* (New York, 1946), pp. 341–42.

propaganda was the fact that the PWB "D" Section found that Sicilians "believed utterly the facts given by Allied propaganda."[17]

ATTEMPTS AT TESTING EFFECTIVENESS

PWB, however, realizing its own lack of experience in making studies of its effectiveness, in late 1943 obtained the services of Professor Stuart Dodd in making some surveys in Sicily. A team of social scientists polled representative samples from the various provinces of Sicily.[18] It was found that about 10 per cent of the 4,000,000 population listened to the radio once a day and that about 64 per cent of the sets were equipped to receive short-wave broadcasts. Of the listening public, most (87 per cent) listened to local stations regularly. Sixty-one per cent listened to London (BBC) an average of eighteen times a month while 48 per cent listened to New York and 24 per cent listened to Algiers twelve and five times a month, respectively. Over 50 per cent listened to North Italy (still under German-Mussolini rule) on an average of every other day. All in all, Allied radio stations "between them enjoy an overwhelming preponderance both as regards the number of those who listen and the hours listened." Lack of electricity was the greatest deterrent to more extensive listening. Few people of those who had radios (11 per cent) listened to German stations, and only 0.1 per cent thought those stations most reliable. Forty-seven per cent of the listeners thought Palermo (under Allied control) most reliable, while 22 per cent thought London most reliable. Only 4 per cent voted for New York and 1 per cent for Algiers, although all Allied stations were giving essentially the same news. There was little faith in Italian stations under German control; 42 per cent voted North Italy least reliable. Of those who listened to North Italy, 66 per cent did so for music, only 7 per cent for news and comment. Ninety-six per cent of Palermo radio listeners heard and enjoyed the "Voice

17 Pollard, p. 78.

18 PWB-AFHQ, *Surveys of Public Opinion Held in Sicily* (Palermo, 1944), originally printed as a "Confidential" document, was given to R. W. van de Velde by Prof. Stuart Dodd. It is now in the Princeton University Library. Hereafter referred to as *Sicily Survey*. All figures and percentages in this and the following paragraph are from this source.

of America" on an average of twenty-one days a month. This program far outweighed the other New York broadcasts in popularity. Only 10 per cent listened to "The Third Front," the next most popular broadcast from New York. In almost all cases the Sicilians heard the New York programs as rebroadcasts from either Palermo, London, or Algiers. It is impossible to say what the listening habits were prior to Allied occupation. It is safe to assume there was some switch-over to Allied stations, but, given the joyous reaction of the Sicilians to the liberation of their island, it would seem fair to say the change was not great.

The surveys found that 61 per cent of the Sicilian people read a newspaper on the average of every other day. It also found that AMG decrees reached urban populations within two days and rural people in about four to five days. Fifty-five per cent of the people learned of these through newspapers, 50 per cent by posters, 44 per cent by word-of-mouth, and 23 per cent by radio. There is, of course, an overlap here with some persons gaining news by two or three means. At that time only Palermo, Catania, Messina, and Caltanissetta had newspapers. All these were under Allied control and all were small owing to shortage of newsprint.

It is unfortunate that this group asked so few questions and that the work was not carried on into Italy. The polling unit was disbanded in January, 1944, in order that an organization along similar lines might be set up in SHAEF.[19] The survey team did serve to verify findings by various less scientific, but perhaps equally reliable, observers that the Sicilians not only were conscious of American propaganda but that they generally accepted it. Since Sicily is one of the most depressed areas of Italy, it is probable that other Italians had even more opportunity to hear and read Allied propaganda. The answer as to whether or not it was as acceptable to the rest of Italy will have to rest on other evidence.

The Psychological Instrument's Use Examined

RADIO

Whenever propaganda is mentioned, radio comes immediately to mind. This is no doubt partly the result of the use made of

19 Letter from Prof. Dodd to R. W. van de Velde, dated May 12, 1953.

that medium by Dr. Goebbels, but it is also a tribute to the extensive use of radio made by the Allies once the war was under way. During the invasion period, the Allies used every conceivable means to saturate the air over Italy with waves of propaganda designed, first, to destroy the morale of the Italians as active Axis partners and, then, as areas became free of fascism and German domination, to restore the Italians' self-respect and transform them into supporters of the Allies. Some emanated directly from New York by short wave, but the bulk of American broadcasts to Italy were either relayed or rebroadcast through the BBC or through stations under PWB control in North Africa, Sicily, and, later, on the mainland of Italy. By early 1943 "The Voice of America" had become a trademark well known to Italians as well as to the rest of the world.[20]

But the mass of American radio propaganda went to Italy as part of the Allied effort through the direct activity of PWB. There were radio stations at Algiers and Tunis. It is interesting and a comment on good planning as well as courage in PWB field units that Radio Palermo and Radio Bari were both seized intact by field teams while fighting still went on in those towns and were immediately set to work broadcasting for the Allies. Because the fighting almost invariably halted electric power in the area, the stations had to be run on power generated by mobile military generators. This cessation of power also meant that the broadcasts were lost on most of the local population who could not use their sets, but it must have made considerable impact on the rest of Italy, which usually knew of the fall of a town first through Allied news announcements. In Bari, for example, the radio was making news announcements eight times a day before the Germans had been cleared from the city. In Naples the retreating Germans destroyed the radio station and PWB teams had to make do with a 1-kw. mobile transmitter which began broadcasting on October 15, just two weeks after that city was entered. By December, 1943, PWB had established a radio station in Sardinia which reached areas in North Italy where Naples, Bari, and

[20] Cf. L. Carlton, "Voice of America: The Overseas Radio Bureau," *Public Opinion Quarterly*, VII (Spring, 1943), 46–54.

Palermo could not be heard. Later a station was set up on the Anzio beachhead which was in constant communication with Italian Partisans in Rome. When Rome fell, in June, 1944, eight additional transmitters were established, some of them for work in the Balkans.

One of the major difficulties faced by the Radio Division was common also to the News Division: the question of linking these various cities so that not only policies and programs could be coordinated, but so that news could be processed rapidly by all the gathering and distributing agencies. Teletype lines were established between Naples and Bari which, by December, 1943, were handling 15,000 words a day for radio newscasts to all of Italy. Palermo was linked by continuous-wave radio. By April, 1944, New York OWI was broadcasting around the clock in Italian, 15 minutes every hour. Some of these programs were for direct consumption, some for relay or rebroadcast. Many were news items for PWB which were incorporated in programs originated in MTO or used in the various papers published by or under the supervision of PWB.

THE PRESS

Printing presses, like radio transmitters, are part of the munitions of war. And PWB teams were also alerted to take them over in all liberated cities and towns. Palermo was one of the first; Catania, Caltanissetta, Messina, Salerno, Bari, Naples, all had their papers as soon after liberation as PWB could manage, and they managed rather remarkably. Palermo fell on July 23; by August 7, *Sicilia Liberata* was being printed on presses made from the wreckage of two Fascist papers left behind by the enemy. At first it was 2,000 copies a week, then it rose to 30,000 a week by the end of August.[21] By December it had a publication size of 40,000 a day. A PWB team, using an old-fashioned hand-operated press ran off 1,500 copies of *Corriere di Salerno* while the building they worked in was under intense enemy fire, only one day after the landing. A day before Naples fell, this paper, now more easily produced, was run off under the head *Giornala di Napoli* to be ready for immediate distribution when that city was entered.

21 Pollard, pp. 84–85.

While this speaks well for PWB work, it should be remembered that in overall figures the total was not staggering. Shortage of newsprint was the principal drawback. Some idea of the per capita dearth of newspapers is indicated by figures for Sicily about three and a half months after the island was cleared of the enemy. Palermo's population of 417,000 had 40,000 copies; Catania's 244,000 had 20,000 copies; Caltanissetta's 50,000 had 8,000; and Messina's 195,000 people had only 5,000 to read.[22] It is remarkable then that the papers had as high a readership as indicated earlier. This fact would tend to substantiate claims by OWI that Europeans were starved for news. The figure of something under 10 per cent for ratio of copies printed to population held true in Naples too, where over 1,000,000 people had only 100,000 daily papers to read.

Both the Radio and News divisions were principally preoccupied with news: truthful news but selected and/or emphasized in such a way that the Allies appeared as sure victors in the war and firm friends of the Italian people. Every effort was made to convince the Italians that Germany was their real enemy. Even though the Campaign of Truth caught the attention of the average American only five years after the war, the propagandists knew the value of truth and habitually stuck to it. This meant admitting the slow progress of the war as well as local reverses. Edward Barrett has stated that the Italians had come to believe VOA (and American propaganda) when it carried General Stilwell's 1943 statement that "we took a hell of a beating." The Allied radio and newspapers attempted to build up their audiences with programs and features which would lighten the fare of straight news and editorial comment, just as those media do here. These were often designed to instruct Italians in a knowledge of Americans and American life. This aspect of propaganda was, however, more notably exploited by other media. Part of the radio time and news space was, of course, devoted to proclamations and announcements of AMG and ACC. Every means of communication was used to inform the people of what the Allies wanted them to do, what new regulations were in effect in regard to

[22] *Sicily Survey*, p. 24.

water use, food rationing, curfew rules, travel and movement, relief, refugee shelters, etc. These and many other matters must be known by a citizenry if they are to remain calm and become a help rather than a hindrance to the occupying troops.

NAPLES—A TEST FOR AMG

The more capably Americans managed these political and economic affairs, the more effective would be their psychological operations. Naples was the first major city on the European continent captured by American troops, and General Clark, knowing the word would spread throughout Italy, was anxious that a good job be done by AMG.[23] The Army engineers restored the water supply, electricity, sewage, telephone, and port facilities. The Italians cooperated. Part of this cooperation was undoubtedly due to the work of PWB in publicizing the friendly attitude of America and its anxiety to help. For example, PWB radio, newspapers, and mobile loudspeaker trucks toured the entire city, warning the inhabitants to leave their homes and shops between certain hours on the day the engineers turned on the electricity. The Germans had left many booby traps and time bombs behind them and it was feared some undiscovered ones might cause great demolitions when the power was restored. Sound trucks were used in all newly captured towns to give immediate instructions to the people, thereby materially lessening the inevitable confusion and fear by civilians subjected to war at first hand. The effectiveness, as propaganda, of American actions in Naples is attested to by the difference in treatment accorded the troops on their arrival in Naples and on their arrival at every Italian city captured from the Germans later. In Naples, when Americans first entered the city, there was no greeting from the natives, there was nobody in the streets. In every other city in Italy which Americans entered they were greeted with flowers and cheers and flags as liberating friends. We had made "good propaganda" in Naples. While the war obviously separated the Italian nation with a "front" across the peninsula, there was always some surreptitious movement by natives across "the lines" at lightly defended places. News traveled with these line-crossers.

23 M. W. Clark, *Calculated Risk* (New York, 1950), p. 213.

AMG also enlisted the cooperation of PWB in the matter of posters. One of the first to be put up on walls, windows, and poles was called *Granai del popolo* (Granary of the People). It was a combination of five photographs and short paragraphs. The first picture showed peasants harvesting wheat with an ox team. The second, showing two smiling peasant women stacking wheat, said, "The harvest has been good, people are filled with hope." The third explained that "men and machines transform the cut wheat to grain." The fourth, showing an oxcart hauling bags of grain, said, "Toward the granary of the people. An equitable distribution means bread for everyone." The fifth, showing men, including an American officer, inspecting stored grain, explained that quantity and quality were checked and that "the Allied officer will advise on the distribution." The object, of course, was to encourage farmers (this was harvest season in Sicily) to bring their grain to the state granary and to assure the people generally that not only was there grain enough for everyone, but that the Allies would see that it was fairly distributed.

Granai del popolo was a popular poster; it was large, it was well illustrated with scenes of natives at work, and it was reassuring. Another early poster was called *Olearii del popolo.* It was not illustrated but carried a large headline in heavy black saying, "Italy has sufficient oil for rationing." Under that it went on to say, "Rationing has been asked for by the dealers. Bringing oil to the refineries of the people [state processing plants for olive oil] is an act of patriotism as well as being required by law. The Government will enforce the collection with full use of the law." Bread and olive oil are two of the most staple items in all Italian households.[24] Lack of other fats causes much wider use of olive oil in Italian kitchens than most Americans realize.

A poster which was used throughout the early campaign and at least beyond the capture of Rome was a simple but effective one. It showed, as the most prominent feature, a German army-

[24] Cf. M. Mead, "Food and Feeding in Occupied Territory," *Public Opinion Quarterly,* VII (Winter, 1943), 618–28, for a discussion of the importance of this aspect of the control of a people.

booted foot lifted as in walking. On the bottom of the heel was the Nazi swastika in iron nails. The boot appeared to be about to be set down crushingly on an outline map of Italy. The only wording was *Ecco il nemico* (Here is the enemy). The idea was to turn enmity from the Allies to the Germans. The Italians were "official enemies" of the United States until the surrender in September, 1943, and defeated "neutrals" until they declared war on Germany in mid-October. They had for years been fed on a propaganda picturing Americans as imperialist, capitalistic vultures and cultural barbarians content to isolate themselves in their rich, materialistic homeland. Not only was it necessary to undo the damage of this kind of propaganda, but it was essential to turn latent hostility and anger at the havoc war was creating in Italy against the Germans and away from the Allies. *Ecco il nemico* was one step in that program.

Another poster designed to place the Italians emotionally on the side of the Allies, which was one done at the request of AMG, said,

> You have helped free Italy. You must now help rebuild the Italy you have so gallantly helped to free. . . . The Allied Military Government desires to leave the administration of the community in the hands of good citizens who are no longer under the yoke of Fascist tyranny.

This poster also served the purpose of enlisting the cooperation of the natives in the administration of their towns and cities, but probably its most telling slant was that it helped restore pride and morale among the Italians who read it. They were obviously earning the respect of these invaders and were not only not despised for having lost the war or hated for having been enemies—they were praised and asked to continue to help. Some posters were used to stress news events of major importance. *Roma e liberata* was one which appeared all over free Italy within a few hours after Rome was taken by the Allies. Literally millions of less spectacular posters containing AMG decrees, and so forth, were affixed to walls and windows all over Italy.

DISPLAYS AND EXHIBITS

Closely related to the poster campaign were the Documentation Centers or "prop shops" of PWB, established at strategic locations in towns and cities. A vacant store usually served as a center where American and British magazines and papers were made available to readers. The walls and windows were decorated with various posters and enlarged pictures. These were sometimes of scenes in America or Britain showing some phase of the war effort on the home fronts of the Allies; they were often news photos dealing with the war. The mounted photos were frequently of scenes taken locally which would bring out what the Allies as a whole were doing for Italy's assistance and reconstruction or what some individual soldier might have done to win the friendship of an Italian child or family.

Theater lobbies were very effective as subsidiary "prop shops." There displays could be used to attract patrons to the theater, which would be showing some American or Allied films, and those who had come specifically to see the movie could study the displays while waiting to enter or on leaving the theater. Actually it was these Documentation Centers which later grew into United States Information Service (USIS) libraries and reading rooms. Even during the war, once some semblance of safety and order were restored to a town, PWB began working out liaison from the "prop shops" with Italian institutions, which soon began asking for specific books or publications on certain subjects. Universities and individual scholars were among the earliest to recognize the possible value of these Allied centers of periodicals and information.

Most posters were printed in MTO; some of the more elaborate ones were shipped from the United States, but much of the paper for posters, like the publications for the "prop shops," had to be sent from America. Some idea not only of the amount of activity carried on by PWB, but of the Army's acceptance of its importance is indicated by the Army's allocation of shipping space to OWI for the period April–November, 1943. OWI New York shipped (in addition to strictly Army supplies to PWB) 900 tons of paper, 180 tons of printing presses and parts, and almost 8,000,000 publications.[25]

[25] House, *Hearings 1944*, pp. 1263–64.

LEAFLETS

A tremendous user of paper was the leaflet program. Most of the leaflets were addressed to German troops and so are of no concern in this study. But many were addressed to Italian civilians. It is impossible to say what effect on Italy was engendered by the leaflets showered on Italian troops while they were still in the war against the Allies. All evidence points to their effectiveness on the Italian soldiers.[26] Most Italian prisoners of war in Tunisia and Sicily were found to be carrying PWB leaflets, usually the "Safe Conduct" surrender leaflet. Pantelleria, a small island off the coast of North Africa, was bombarded with leaflets as often as bombs and was said by some to be the only place captured by Allied air power. Whole units on Sicily surrendered to approaching Allied troops without firing a shot, especially in the west end of the island where there were no German troops. These events are important to this study only because they show one of two things or a combination of them. Either all Italians wanted an end to the war and therefore home-front lack of morale helped break that of the troops, or troop morale was so low that it began to be influential on the people back home. In either case it seems fair to say that PWB leaflets had a part in the destruction of Italy's will to continue the fight as part of the Axis.

After the fall of Tunis strategic leaflets were aimed more and more at the Italian home front. By August, 1943, PWB in Tunis was producing about 7,000,000 a week.[27] Many were addressed to the Italian home-front worker, urging him to slow down, to avoid helping Germany, worrying him on the subject of growing Allied and shrinking German air power, which meant an increasing amount of bombing on Italian war-goods

[26] Cf. Barrett, p. 37; U.S. Congress (Senate), *Hearings on First Supplemental National Defense Appropriations Bill for 1944* (Washington, 1943), p. 18; House, *Hearings 1944*, p. 1265. Pollard, pp. 53, 55, 77, tells of the answer of the Italian Commander of Pantelleria when asked if leaflets had had any effect. The Italian general "pulled one from his pocket and described how his men had knelt in prayer in front of his headquarters, asking for surrender." Page 80 indicates that in a sample of the first 300 Italian POW's taken in Sicily all had seen and believed PWB leaflets.

[27] Pollard, p. 79. By December 1943, 15,000,000 leaflets a week were dropped on that part of Italy in German hands (*ibid.*, p. 102).

factories and communications facilities. Italians were warned away from railyards and seaports. The Eisenhower proclamation at the time of the Sicilian invasion was, in addition to radio, put out in millions of leaflets which were showered down all over Italy.

ADVANTAGES OF LEAFLETS OVER RADIO

Leaflets were a constant supplement to propaganda put out by radio. Night bombers, flying low over a city, could fairly well blanket it with the "paper bullets." They lack the speed and flexibility of radio, but they also can get behind enemy lines and have two advantages radio lacks. They are more permanent, can be reread time and again by many different people, and they are visual media and, therefore, lend themselves to graphics such as pictures, cartoons, and caricatures. The latter were especially telling in the anti-Mussolini campaign prior to his fall. His face was ideal for the caricaturist and his position vis-à-vis Hitler was exploited to belittle and demean him in Italian eyes. There is another possible advantage a leaflet has over radio: it is a written document, and, therefore, a promise made by leaflet may have more weight than one made over the air waves. A possible proof of this feeling was the fact that the original "Safe Conduct" leaflets used early in the Tunisian campaign were unsigned and experience proved that prisoners would have had greater faith in them if they had borne some "official" signature. This was done and the results were noticeable. It is therefore possible that Italians receiving "written" promises of future freedom to choose their own government, of the return of Italian POW's if Allied prisoners were not sent to Germany and were helped to escape, etc., were more inclined to believe them than if these promises had been made only by radio.[28]

Generally the strategic leaflets addressed to the civil population were prepared at a level above a numbered army, but General Clark, commanding the U.S. Fifth Army, several times made specific requests for PWB leaflets to further the tactics of the combat units. One such was a leaflet, in Italian and Ger-

[28] Cf. Butcher, pp. 368–69; Pollard, pp. 82–83.

man, addressed to the inhabitants of Naples as well as its military occupiers. One hundred and fifty thousand leaflets in each language were showered on the city, calling for its surrender and promising "large stocks of food" when the Allies entered. He again ordered leaflets when he wanted to warn civilians to evacuate the Monte Cassino Abbey before it was to be bombed. These leaflets served the double purpose of showing the Allies to be conscious of the dangers of war to innocent people and of attempting to justify the bombings and avoiding their being used as Axis propaganda "proof" of American barbarism.

AFHQ had perhaps set the pattern for this type of prior explanation-warning leaflet. On July 3, 1943, even before the Sicilian invasion, leaflets had been dropped on Rome warning that the railyards would be bombed if the Germans continued to use them. The leaflet warning was repeated on July 18. On the nineteenth the yards were bombed in broad daylight. Six minutes later VOA interrupted all European broadcasts to announce the raid, the reasons for it, and to emphasize the fact that great care was taken to avoid the Vatican and all cultural places. Some students of the subject place sufficient stress on these warnings, followed by daylight raids coupled with the invasion of Sicily, to infer their decisiveness in Mussolini's fall.[29] They do admit that other factors played a part, as they must have, but these were indeed momentous events and strong propaganda fare for the Italians to mull over. In any event, these serve as examples of a specific psychological operation's being integrated with a specific use of military force.

PERIODICALS

Booklets of varying length and content and periodicals were also used and may have been a more lasting, if less spectacular, kind of propaganda. Both were distributed to Italians in liberated areas, usually through "prop shops," but also at any gathering of the populace, such as at movies. The two periodicals were *U.S.A.* and *Victory*. *U.S.A.* was a monthly publication of the digest type, running about seventy pages and con-

[29] Cf. Pollard and Barrett.

taining about sixteen pages of pictures. It contained articles condensed from American magazines, newspapers, and books which were believed to have lasting interest and would interpret for foreigners something of American thought and customs, as well as show the vigor of the American war effort. *Victory*, published every two months, contained about eighty pages, including twelve pages in color pictures. This was original work in that the articles were all prepared by or for OWI. In an effort to make this appear like other magazines it contained advertisements of the "institutional" type which were paid for by the advertiser. Both magazines were sold at a nominal price with the idea that what is paid for will be regarded more highly and perhaps shown to others with pride. Both were published in the United States in various languages[30] and shipped abroad. Because this took time, the articles had to have a timeless rather than a timely quality.[31] This also made it more probable that they would be kept by the purchaser and shown to friends from time to time. They were very popular, often being sold out before would-be purchasers had reached the head of a queue; in some areas a black market in these magazines sprang up.

BOOKLETS

In the words of an OWI report to Congress, the booklets each had "a specific propaganda purpose and objective . . . usually directed to a specific country or group within a country." They were usually produced in the United States under the supervision of a regional expert and often at the suggestion of men in the field (PWB, for those written in Italian). Elaborate ones were sold at a nominal price, but most of them were distributed free to a selected mailing list or in answer to a request or

30 *U.S.A.* was printed in 75,000 Italian-language copies and *Victory* in 125,000. House, *Hearings 1945*, pp. 149–52.

31 Sometimes as much as six months elapsed between "deadline" and distribution because of priorities of shipping space and the fact that the magazines were made up in advance of actual occupation of an area. House, *Hearings 1945*, p. 138.

given to visitors to "prop shops." Some of the booklets are worthy of being considered separately.[32]

Italian American Friendship was published in Italian under the title *Agli italiani.*[33] It was a small (12-page) booklet bearing a reproduction of the Italian hero Garibaldi on the cover. The opening paragraph said,

We wish that you Italians accept this message which comes to you from the United States of America[34] in the spirit in which we send it: as a salute and a promise of aid; as an expression of the spirit of brotherhood we have for you; a natural sentiment between peoples of two countries which have inherited an old friendship.

32 The tremendous size of the booklet program may be seen by the following partial list:

Booklet	Date	No. of Copies in Italian
	1943	
Italian-American	June	2,000,000
Building a Ship	July	100,000
185,000 Warplanes	July	25,000
What Is Happening to Italian POW's	August	1,500,000
Atlantic Charter	August	5,000
Facts on the U.S.A.	August	500,000
Sicilians of American Origin	August	50,000
Conquering the U-Boat	September	150,000
War Messages of FDR	November	7,500
Wings of America	November	50,000
U.S. Coast Guard	December	100,000
Luigi Lucioni	December	100,000
Aircraft Carriers	December	100,000
U.S. Pacific Fleet	December	100,000
Italian-American Friendship	December	100,000
Pocket Calendar (1944)	December	250,000
The Master Race	December	100,000
	1944	
Battle for Africa	January	100,000
Moscow Conference	January	100,000
Women of the U.S.A.	January	100,000
UNRRA Conference	January	100,000
Italy Fights with the UN	January	100,000
Alphabet Primer	January	100,000
U.S. Army Insignia	February	200,000
Joseph Stella	February	100,000
Umberto Romano	February	100,000
Ernest Pellegrini	March	100,000
This Is the U.S.A.	April	50,000
Documents on Liberty	April	20,000
War in the Pacific	April	150,000

33 See n. 32, above. It is difficult to be sure that the booklets named in the report to Congress are always precisely those which the present writer has studied. For obvious reasons OWI did not date the booklets and frequently the Italian title is not a direct translation of the American one. The two have to be "matched" by a study of content and references in the text which hint at the approximate date of publication.

34 In booklets, periodicals, and films the United States rather than the Allies could be stressed even though they naturally were never used to undermine Allied solidarity.

The booklet speaks of the millions of Italians who have emigrated to the United States and their part in building America. A link is suggested between Jefferson's writing of the Declaration of Independence and Mazzei, a Tuscan whose writings the American had studied, and Italy's own *Risorgimento.* It is suggested that Italy stands on the threshold of a rebirth and that the promises of the United Nations at the Moscow Conference (October, 1943) prove the resolve of those nations to help Italy destroy all traces of Fascist tyranny. Freedom of speech, press, religion, opinion, and association will be restored in Italy through the aid of the Allies. The whole gamut of Italian-American relations from Columbus and Vespucci, through Garibaldi's sojourn in the United States, to American love of Italian music and art, is covered. Thanks is given for the kindness shown by Sicilians to wounded Americans. "The U.S. has fought the Fascist regime; it has never considered itself at war against the Italian people," says the booklet which goes on to quote from a speech by Matteotti, the famous Socialist deputy assassinated by the Fascists.

There are several themes here. One is the cultural and racial ties between the two peoples. This theme was carried on in other booklets too.[35] It was often "documented" in person by the great numbers of Italo-American soldiers who could and did communicate with the natives in Italy in their own language, sometimes complete with regional accents and even *patois.* Another theme is that of Americans fighting for Italy, which was also carried in various other booklets.[36] The theme of

[35] *The Alphabet Primer,* called in Italian *L'abc Italo-Americano,* which was cleverly illustrated, carried such statements as, "A is for America. The U.S.A. is the home of 6,000,000 persons of Italian origin." This book also played on the Garibaldi–pro-American theme. The series dealing with American artists of Italian origin, each one a very brief biographical sketch with pictures of artists, such as Luigi Lucioni, Umberto Romano, Jon Corbino, Ernest Pellegrini, all played up the cultural links existing.

[36] *L'abc Italo-Americano* said, "B is for *Battelli da sbarco* [landing boats] which brought American soldiers to Italy." *Come gli Americani di origine italiana combattono per un'Italia libera* was an illustrated booklet which showed many prominent Italo-Americans such as C. Poletti (former lieutenant governor of New York, member of AFHQ Civil Affairs Division), Daniel Noce (American general), F. La Guardia, G. Salvemini, M. Ascoli, etc. It stated there were 600,000 American soldiers of Italian ancestry. The good treatment accorded Italian POW's and interned merchant seamen was brought out. Nine

antifascism and return to democracy was made appealing to Italians partly by reference to Garibaldi and other great names in Italian history. Here again *L'abc,* which dealt with every letter in the Italian alphabet and could cover a range of subjects, stressed this theme, saying, "R is for resistance and rebellion of the Italian people against Fascism." Along these lines were the booklets *Why We Fight* and *Documents on Liberty.*[37]

The 1944 *Calendar,* which was published in a compact folding form so that it could be carried in pocket or wallet, was another means of covering a wide assortment of themes with attention-holding brevity. It contained short quotes from Toscanini, a man much loved by Italians and highly respected, on liberty and Italy and from President Roosevelt on "decency and humanity." There were pictures of American bread being distributed in Sicily; of American troops being happily received by Sicilian townspeople; of American GI's making friends with children in Italy.

The general joy which Americans felt when Italy declared war on the Axis was shown in a booklet, *L'Italia si stacca dall'Asse!* This issue was almost entirely pictures: of Italo-Americans putting up flags of both nations; of headlines of Italian-language newspapers in the United States on October 13, when the news of cobelligerency broke; of Italian POW's in the United States beering and cheering in PX's at their camps. *Combattenti degli Stati Uniti,* an illustrated, fifteen-page booklet, showed American soldiers with their names, giv-

---

hundred of these last were released by the United States on October 2, 1943, which, when duly publicized, gave great credence to the statements such as those contained in this booklet.

37 *Per che cosa combattiamo* was a large booklet with bright red covers. It contained a collection of excerpts of speeches by Allied war leaders, FDR, Churchill, etc., as well as excerpts from the Atlantic Charter and Allied conferences. It said, "These words . . . make a part of the history of our time—a history in which patriotic Italians, through their resistance to fascism and nazism, have written an heroic page." *Documenti della libertà* was a "pocket size" issue containing various historic documents such as the Mayflower Pact, Patrick Henry's famous "Give Me Liberty" speech, the Declaration of Independence, Virginia's Statute on religious freedom, the Preamble and the Bill of Rights of the U.S. Constitution, the Four Freedoms, etc. Each was accompanied by a short editorial comment explaining the circumstances leading to the document and its importance.

ing certain biographical background, such as national ancestry (including Italian, Swedish, Belgian, Australian, French, Greek, American Indian, etc.) and the different regions, educational levels, and occupations from which the men came to join the armed forces. *Ali d'America* was a cartoon-type booklet, in the manner of Ripley's *Believe It or Not,* whose ostensible purpose was to give the silhouettes of American planes with identifying data. It also showed, in a cartoon series, the history of American aviation from the Wright brothers to World War II. The fact that the United States alone was producing 12,000 planes a month was stressed.

Booklets often sought to bring home to the Italians the strength and determination of the United States, as a residue of that earlier propaganda over the air and in leaflet form which exploited this when there was little else to be said. In January, 1944, an illustrated, fifty-page booklet, *La battaglia de l'Africa,* outlined the start of the African conflict with the war of "Fascist aggression" in Ethiopia.[38] It told, and illustrated with maps, of the seesaw battles of North Africa, and, while calling the battle of El Alamein one of the decisive battles of history and lauding General Montgomery, it saved its impact for *arrivano gli Yanke!* in the "greatest armada in history." It spoke of the "overwhelming'" power of the Allies in the invasion of Sicily in 3,000 ships and craft. Several pages were devoted to the reconstruction undertaken by the Allies and to the arrival of food and other relief supplies. An earlier booklet, *Building a Ship,*[39] gave in brief, illustrated form the story of how Kaiser had produced a merchant ship, from keel-laying to launching, in less than five days. It stated that America was now turning out ships at the rate of five a day; that over one hundred had been launched in March, 1943, alone.

There were, of course, many OWI pamphlets which sought to enlarge the understanding of Italians for America. Elmer Davis had testified to Congress, "We are confident enough in the quality of American life to believe that on the whole, the more foreign peoples know about us, the more highly they will

[38] Perhaps a propaganda mistake as most Italians "went along" with that war.

[39] Released in Italian under the title *Una nave costruita in 4 giorni, 15 ore e 25 minuti.*

respect us." That policy was the basis of such booklets as *La Musica negli Stati Uniti,* which was designed to show in words and pictures all types of American music and the maturity and breadth of American culture. At least two, *Panorama degli Stati Uniti* and *Qualche notizia sugli Stati Uniti,*[40] dealt in a sweeping hand with various aspects of America. As might be expected, *Panorama* was illustrated. The pictures were of a multitude of subjects. Each region—the West Coast, the Southwest, New England, etc.—was illustrated with excellent photographs. There was a balance of agricultural scenery and scenes of industry. Faces of Americans—workers, students, housewives, cowboys—were well presented. This was an elaborate work and may have been one of those sold. On the other hand, *Facts* was a small booklet which presented basic facts in terse sentences, "The U.S.A. is a free republic. Its citizens vote, have freedom of speech and religion. . . . The U.S. is the fifth largest nation of the world in territory. Brazil, Canada, China and the U.S.S.R. are larger. . . .[41] Under American law all men are considered equal." This work also contained pictures of great assortment, but all of American flavor: the Capitol dome, the Statue of Liberty, a traffic "clover leaf" from the air, New York City at night, steel mills at night, FDR addressing Congress, Yellowstone National Park, prairie farm landscape, etc.

At least two booklets were published which dealt with American government. One[42] explained in simple language the divisions between the executive, legislative, and judicial branches of the federal government. It also explained something of the relationship between states and the nation, as well as the major political parties and the relation of pressure groups to all. It concludes that "the U.S. has proven that efficient government is not incompatible with freedom of citi-

40 From an examination of the texts these were probably referred to in OWI report to Congress as *This Is the U.S.A.* and *Facts on the U.S.A.*, respectively.

41 The mention of these four larger nations by name was probably done on purpose, since all were fighting on the same side as the United States. To thoughtful Italians it must have appeared a tremendous array of power. Note that this booklet was published before the Italian surrender.

42 *Come funziona il governo degli Stati Uniti* was "official" looking, as it bore the seal of the United States on its cover.

zens." The other attempted the dual role of reassuring the Italians that the 1944 election results, no matter who won, would not sway America from her determination to fight the war to a successful finish, and to use the fact of a national election in the midst of war as an illustration of a viable democracy at work. There were many other booklets, but these will suffice to show the propaganda effort and generally the appeals used in their makeup. Their popularity has already been attested; one observer was quoted as saying that the arrival of a PWB sound truck which gave out newspapers and booklets as it passed through a liberated town " 'was like a blood transfusion in this intellectually sick, bored and hopeless body of people.' "[43]

THE MOVIES

The other great propaganda medium which PWB exploited during this period in Italy was the movies. Three types of films were used: regular commercial "feature" pictures, newsreels, and documentaries. Motion picture theaters and all existing stocks of films were seized by PWB immediately after a town was taken by the Allies. The films were carefully censored to eliminate any which were of Axis, anti-Allies vintage and those American or other films believed to portray America or her allies in an unfavorable light.[44] Aside from the long-range value of films as propaganda, the Army very early found them of great help in keeping natives calm and contented. The Army usually had to furnish the necessary electricity for the showings. This was done in Palermo, for example, on August 14, when the first theater was opened there under AMG-PWB auspices. Although patrons had to pay for seats, on the theories that what is paid for is appreciated and that a paid-for show looks less like propaganda, the initial performance was sold out before the start of the show. The first movie house was opened in Rome just six days after the city's capture and a year later 120 theaters were operating in the capital. At the time of the invasion 7,500 reels of "the very best" American

[43] Quoted in Pollard, p. 123.

[44] An interesting sidelight on how hard it was for OWI to convince certain congressmen of the value of movies as a part of the war effort is to be found in House, *Hearings 1945*, p. 165.

films were taken to Italy for showing. This represented about forty full-length pictures, whose principal role was to furnish the lure to attract paying audiences to whom OWI newsreels and documentaries could then be shown without the entire performance being discounted as "mere propaganda." These feature pictures were donated to OWI "for the duration" by the different American film companies.

The shorts, both newsreels and documentaries, were usually made by OWI. These were made both on standard 35 mm. film and on 16 mm. film for showing on PWB mobile equipment to clubs, schools, and other institutions, or in village squares sufficiently removed to be relatively safe from night air attacks by the Germans. The newsreels were the usual composite of many photographers working in all parts of the world. OWI London assembled them and shipped them to PWB. The latter had usually dispatched to London shots of current events from the fighting front in Italy but a regular local addition featured scenes in the liberated area; the arrival of Allied relief, the cooperation of the populace with AMG in reconstruction or rehabilitation, and so forth. Many such sequences were made at the request of AMG. After mid-1944 much of this work was done by PWB itself in its newly created film "studios" in Naples. The newsreel was officially known as *News of the Free World*, but OWI reports refer to it as *United Newsreel.*

Most of the documentaries were short, one or two reels—a few were longer. Most of them were made by OWI in the United States,[45] but by July, 1944, PWB was making a few in Italy at the request of ACC. *Swedes in America* starred Ingrid Bergman, already a favorite with Italians, and showed the assimilation of a group of immigrants into American life. *Cowboy* attempted to show that man as he really was, a rugged, hard-working fellow who did not spend most of his time shooting other men, the movies notwithstanding. While possibly deglamorizing the man of the plains, it was a good vehicle for showing the vastness of the American West and its productiv-

[45] House, *Hearings 1945*, p. 162. Among those completed by April, 1944, were included *Swedes in America* (2 reels); *Cowboy* (2 reels); *A Journey* (2 reels); *Autobiography of a Jeep* (1 reel); *The Town* (2 reels); *City Harvest* (1 reel); *People to People* (3 reels); *Steel Town* (2 reels); *Hymn of Nations* (4 reels); *Yellow Springs* (1 reel); *Valley of the Tennessee* (3 reels).

ity and organization. *Autobiography of a Jeep,* that ubiquitous vehicle well known and admired by Italians for its stamina and versatility, showed the entire production line process and was impressive to Italians, especially of southern Italy, who had never seen such large-scale industrial skill and machinery at work. *Hymn of Nations* was a musical composite of Verdi's *Hymn of the Nations* and the national anthems of the United States, Great Britain, France, and Italy. Its major purpose was to show Italians that their beloved Toscanini (who directed the orchestra in the film) was on the side of the United Nations. It included some documentation on various exiles from fascism, including Toscanini himself. This was a most popular film with Italians. *The Valley of the Tennessee* not only served to stress the tremendous size and wealth and potential of the United States, it was a classic of technical instruction by use of movies. Land reclamation was the underlying theme, and this has been and remains one of Italy's great problems. It could not fail to make a lasting impression on all Italians. There were also films on care and feeding of children, prevention and care of diseases, rural electrification, and others directly designed as instruction but having an indirect propaganda value.

Among the documentaries produced in Italy by PWB, using local talent as actors and familiar scenery as "location," were such titles as the two-reel *Granai del popolo.* This was conceived to serve the same purpose as the similarly titled poster, but could be made considerably more effective in motion picture form. In addition to pointing out the necessity for peasant and townsman to cooperate to stamp out the evils of the black market, it presented brief interviews with members of the Bonomi Cabinet, many of whom were unknown to most Italians. Another locally produced "short" was *Fosse Ardeatino,* which told the full story of the infamous massacre in those caves, rumors of which had long since stirred up great hate against the Germans and their Fascist lackeys.

THE OVERALL EFFECT

This rapid summary of the work of PWB in its relations with the Italian people can hardly describe the total effect of the day after day, week after week, month after month impact of a really tremendous propaganda program. Only those closely connected with it, either in its production or in its reception, are able to realize the ceaselessness and the scope of the operation. In some ways less spectacular than the U.S. psychological operation which was to be undertaken at the time of the 1948 elections, it was, because circumstances permitted it, far more all-embracing and all-pervading. It was the first time that the United States had had the inclination or the opportunity to practice all-out propaganda on a nation; it was to set the pattern and reveal avenues of success and pitfalls of error to those who later turned their attention to Germany.[46] But, as has so often been said, propaganda in its spoken or written form is useless or worse than useless unless coordinated with policy and action.

What were U.S. policies and actions in Italy which either strengthened our propaganda or undermined it? It will be impossible to examine them all, but certain outstanding ones must be reviewed in order to form an opinion on the overall effectiveness of American psychological operations during this period, and on its lasting quality. In a material sense both American policies and actions were on the positive side. Vast quantities of food, medicine, clothing, and other relief supplies were brought to Sicily and Italy. Reconstruction of roads, rail lines, power lines, and port facilities, while limited to those required to assist the military effort, were undertaken in many areas and on a large scale. Italian labor was put to work and paid for it. The lira was stabilized, artificially to be sure. But these were the first steps in restoring a shattered and chaotic economy. Public services, such as water, sewer, and electric, were repaired or improved, partly to facilitate the business of the military, partly to assure a relatively quiet and cooperative population at the rear of the combat forces. Health measures, again a mixture of self-protection and humanitarianism, were instituted. Thus whole towns and whole populations were dusted against disease-bearing (particularly typhus) lice. In addition to this physical evidence of aid, there were the count-

[46] Cf. letter by General Eisenhower facing frontispiece in *The Psychological Warfare Division, Supreme Headquarters, Allied Expeditionary Force* (Bad Homburg, Germany, 1945), booklet; loaned to R. W. van de Velde by Edward W. Barrett, former Assistant Secretary of State for Public Affairs.

less personal gifts and kindnesses given by a naturally generous American soldiery to destitute Italians and especially to the appealing, dark-eyed children.

Political Blunders: Negative Propaganda

It is sad perhaps, but appears to be true, that the positive effect of all these material things can be quickly nullified by some political blunders in what might be called the spiritual field. Much has been written about the political policies, or lack of them, of the United States toward Italy during the period under discussion. The complaints raised by those who claim the United States blundered fall roughly into five generalizations: (1) the use of local Fascist officials by AMG and ACC; (2) support of the House of Savoy and the Badoglio regime; (3) generally avoiding or ignoring the anti-Fascist political leaders who returned from exile; (4) generally ignoring the liberal elements, particularly labor, which had remained in Italy; (5) rebuffing the Partisans, especially those of North Italy who did materially contribute to the success of the Allied campaign. It can be seen at a glance that all are related and all stem from the same factors. Without going deeply into causes, or the right or wrong of the matter, some comment must be made on the psychological effect on the Italian people of all five. The problem is not whether or not the generalized charges are true or, if true, whether or not the policies were justified by reasons of necessity; the problem arises rather from the fact that they were *believed true* and that they were *believed unnecessary* by many Italians (as well as many Americans).

Charge No. 1 is largely true, as any observer at the time will testify. After twenty years of Fascist bureaucracy, there were available almost no civil servants on any level of government who were not Fascists by reason of conviction or by reason of necessity in retaining a job. Without a tremendously multiplied civil affairs staff, the Allies could not have exercised a reasonably operative control of the local governments without using many who had served fascism and had thus become tainted with it. The Allies recognized this fact and issued orders to use only those persons who were not "active" party members.[47] This would naturally eliminate all prominent Fascists, but at most local levels both prominence and party activity were necessarily always limited anyhow. AMG's orders to abolish the Fascist party and its influence were probably as well carried out as circumstances permitted, but what the citizen of the village or small town saw was that, with the possible exception of the mayor, the same people were running the municipality. The principal change was the substitution of a military Town Major for the previous Fascist regional overseer. Politics in Italy is largely a matter of personalities and to the provincial Italian, whose political horizon is often limited, the personalities remained unchanged; therefore, Allied talk about destroying all traces of fascism was obviously discredited. This charge by itself might have been mitigated if it had stood alone and especially if known liberal leaders had been on hand to explain the necessities of the case to their countrymen. Its real impact was made only by the addition of other charges.

Charge No. 2, support of the House of Savoy, was probably the most widely criticized political action taken by the Allies in the entire Italian campaign.[48] There is ironically much evidence that Prime Minister Churchill, rather than anyone in the United States, was responsible for this decision, but General Eisenhower recommended supporting the king and the Badoglio government.[49] The AFHQ commander who authorized the "Darlan deal" as a matter of military expediency felt that setting aside the existing regime in Italy would have been too great a strain on the available manpower which he thought, perhaps correctly, would then have been needed in great numbers to control the civil affairs of Italy. Whatever the reasons

[47] O. K. D. Ringwood and C. G. Haines, "AMG in Sicily," *Foreign Policy Reporter*, XIX (October 1, 1943), 180. Cf. M. H. H. Macartney, *The Rebuilding of Italy* (Cambridge, 1945), p. 29.

[48] Cf. Ringwood and Haines, p. 180; G. A. Borgese, "Commemoration of Fascism," *Atlantic*, CLXXV (February, 1945), 68–74, *passim;* Sevareid, pp. 384–85, 424; E. A. Mowrer, *The Nightmare of American Foreign Policy* (New York, 1948), pp. 129, 135; and PWB Directives from August 30 through November 1, 1943, to avoid mention of the House of Savoy and to "assume" the majority of Italians to be rallying to the support of Badoglio, without causing the Allies to appear to side too closely with his regime, indicate the ticklishness of the situation.

[49] Butcher, p. 422.

and no matter how sound the military estimates, it is certain that to those Italians who had held fast to democracy all through Mussolini's years of power this was a great blow. For example, both Croce and Sforza wanted the immediate abdication of the king in favor of his son, Umberto, while northern leaders like Parri wanted a democratic republic. Both groups wanted a broader-based government. Badoglio's Cabinet contained some of Mussolini's ministers, "but no representative of the six underground democratic parties of the north."[50] The Allied insistence on retention of the king and Badoglio appeared to Italians to be a sign of mistrust of the Italians themselves.[51] This did not match the fine words, so often spoken, that the war was against fascism, not against Italians. Not until April of 1944 did Italian public sentiment succeed in making the Allies cause Badoglio to broaden his Cabinet. Long before that, it had become evident to the ordinary Italian that the Allies, particularly the British, were fearful that only the existing regime could keep Italians from one another's throats. The Committee of National Liberation delivered a letter to ACC for delivery to Churchill saying Italians were "surprised and sad"[52] at his attitude. Some of this resentment was bound to rub off on Britain's ally, America.

Charge No. 3, ignoring returned exiles and other obviously anti-Fascist political figures, has been partially substantiated in the foregoing paragraph. Croce, Sforza, Parri, Sturzo, and many others, were made to feel not only impotent in the formation of plans for Italy's future, but unwanted in current operations of government. These were national, actually international, figures. The feeling that the Allies, despite their high-sounding phrases, were the supporters of reaction was multiplied many fold on lower levels where they also frequently shunted aside men who approached them, declaring their anti-Fascist records and offering their services. Churchill's *Closing the Ring* explains why he felt as he did about the prominent men. The problem at the lower level was created in part by the ignorance of the civil affairs officer concerned, partly by the

50 H. K. Smith, *The State of Europe* (New York, 1951), p. 216.

51 Cf. N. S. Jucker, "Italy: East or West?" *Political Quarterly*, XXII (April–June, 1951), 175–85.

52 *New York Times*, February 29, 1944.

difficulty of sorting out the true anti-Fascist patriot from the mass of turncoat opportunists who attempted to climb on the new band wagon, and partly as a result of AFHQ's order prohibiting AMG from having "political implications of any kind."[53] Although it is patently impossible for a group such as AMG, whose purpose is government, to be completely unpolitical, the order served to force AMG and ACC officers to avoid what appeared to be clashing political forces. The easiest way to do this was to accept the incumbents in office, except where they were glaringly unacceptable to an articulate majority. With their leaders gone and their regime in defeat and disgrace, most Fascists were content to play the quiet, innocent game; they were docile and generally obedient to Allied orders. The real anti-Fascists, on the other hand, were vociferous in their demands for change and, unfortunately, therefore appeared to the overworked, confused American officers, particularly of the AMG, as "trouble-makers" to be avoided if possible.

The Allied avoidance of indigenous liberal elements has been partially covered in the preceding review relative to the lesser lights of antifascism. AMG, under polices from AFHQ, did eliminate the syndicates and the corporations as parts of the Fascist regime. In this they helped to free Italian labor from the yoke Mussolini had placed upon it.[54] When free workers' organizations sprang up in Naples early in 1944, two organizations came into being, one largely Catholic, the other largely Communist. It had been the North Italian workers who had been largely responsible for the fall of Mussolini, but Mr. Churchill's fear of "bolshevism" in Italy and Americans' general fear of communism being as great as it is blind, led to an avoidance of "doing business" with the more militant labor leaders of the Left, Communist and non-Communist alike. It has been claimed that AMG "forgot" labor and that even after all of Italy had been liberated there were only ten "labor officers" in Italy.[55] Again, whatever the causes, large masses of

53 Ringwood and Haines, p. 180.

54 C. G. Haines, "What Future for Italy," *Foreign Policy Reports,* XIX (October 1, 1943), 176.

55 Cf. M. F. Neufeld, "The Failure of AMG in Italy," *Public Administration Review,* VI (Spring, 1946), 137–48. It seems more likely to the present writers that the military command "sidestepped" rather than "forgot." Military men

Italians who believed themselves progressive and knew themselves anti-Fascist found their advice unwanted and their problems of little concern to the Allies.

The rebuffing of the Partisans of the North, whom close observers said were the "best and bravest . . . in all Western Europe"[56] will be dealt with in chapter vi. It is mentioned here merely to place it in the classification of Allied actions which appeared to many Italians to detract from the luster of hope which Allied propaganda had promised and from the feeling of good will which widespread aid and relief measures had created.

---

are generally conservative, believers in the status quo and in order. They do not, as a rule, understand the problems of American labor and are apt to decry strikes as violations of order. How much more complicated and explosive must the Italian labor situation have seemed, where politics and political ideology are so closely interwoven with the other factors, such as wages and hours, which are more familiar on the American labor scene. This would seem to be an area in which the State Department should have given greater guidance to the military.

[56] H. K. Smith, p. 217. Smith quotes E. Wiskemann, who worked with both the French and Italian undergrounds, as saying there was less collaboration with Nazis in Italy than in France.

• CHAPTER VI

American Psychological Operations during the 1948 Italian Elections

By the opening of the campaign for the 1948 parliamentary elections, Italy had once more become an independent, sovereign state, at least in the international law sense of that term. She was far from independent economically, but Allied troops had been withdrawn in December, 1947, under the terms of the treaty of peace, and the new constitution of the Italian Republic came into being on January 1, 1948.

It is difficult to point to any specific date as the start of the political campaign of that year. In many ways it might be said to have begun even long before the final liberation; certainly it was well under way before the new Republic was official. Indications of what was to come in the later campaign were surely in evidence by the time of the 1946 referendum and the Constituent Assembly election.[1] American aid and propaganda were also evident throughout late 1946 and 1947, but the crescendo of activity and emotion reached in April, 1948, can perhaps be said to have commenced in that year.

Political Developments Prior to the Elections

In order to understand the fever pitch of the April elections, however, it will be necessary to give a brief background of the major political events and the situation which preceded it. And

1 Held on June 2, 1946. As early as mid-May, 1946, certain Catholic bishops were warning their communicants to vote against Marxist and anticlerical parties.

in order to attempt an evaluation of American influence in these elections, it will be necessary to enumerate some of the more spectacular psychological operations undertaken earlier. Fortunately for the student in this field, an Italian polling organization was founded in 1946 and data for a study of some opinions and opinion trends exist for the years 1947 and 1948.[2] It should be borne in mind that the many U.S. activities, both in deed and word, governmental and private, which were directed at Italy were designed to keep friendly and cooperative Italian policy-makers in power and to exclude from power men who would certainly *not* be friendly to the United States and the West. It should also be remembered, however, that these actions were largely the result of the intuition and common sense on the part of real and self-appointed experts on Italy. If they were successful, the credit belongs to those individuals and to a fortuitous set of other circumstances. There was no real "plan"; there was only haphazard intelligence; indeed, there was very little continuing policy because the objective itself was limited and of short duration. There is no evidence of a truly long-range objective for Italy other than a vaguely articulated one of keeping Italy from "going Communist."

THE MONARCHY IS DEPOSED AND A CONSTITUENT ASSEMBLY ELECTED

One significant political event preceding the 1948 campaign which had a bearing on it was the referendum on the question of monarchy versus republic. This was a heated issue and a rather close one. The final vote on this was 12,717,923 for a republic and 10,719,284 for a monarchy. The old North-South division was very evident here. The North voted by a four million majority to abolish "the monarchy that had licked Mussolini's boots for twenty-five years."[3] The South voted by

2 Prof. P. L. Fegiz of Trieste University, with a group of scientists and businessmen, founded, in Milan, the *Istituto per le ricerche statistiche e l'analisi dell'opinione pubblica.* This organization, called DOXA, has achieved international recognition and has collaborated with the Gallup organization and others.

3 H. K. Smith, *The State of Europe* (New York, 1951), p. 213.

a two million majority to keep it. When the results were known, monarchists in the South demonstrated and, in Naples, violence resulted in eleven killed and fifty wounded. Nevertheless, Umberto, who had held the throne only thirty-five days, relinquished it and went into exile. Once this had been done, the issue calmed considerably and, despite the existence of a Monarchist party, the question of a monarchy was not vital in 1948. But the closeness of the overall vote should not be lost sight of.

Concurrent in time and probably greater in importance than the referendum was the vote for a Constituent Assembly of 556 seats. Over 88 per cent of the eligible voted. The Christian Democrats emerged as the strongest party with 207 seats, the Socialists were second with 115, and the Communists third with 104.[4]

ALLIED BLUNDERS—POLITICAL AND PSYCHOLOGICAL

The first tragic result of Allied treatment of all "leftists" was the utter defeat and dissolution of the Action party in this election. Some of its members sought benediction in the ranks of the Christian Democrats, others, more firm, felt that only by supporting the better-organized and apparently stronger parties further to the left could any liberal program be undertaken. Probably particularly harmful to the Action party was the fact that its leader, Ferrucio Parri, was the first Prime Minister after the end of actual hostilities. He followed Bonomi, who resigned in July, 1945, being unable to survive the liberation of North Italy. As it was, the Allies failed to turn over the areas of the North to Italian civil control in October, 1945, as promised, and it was Parri who bore the blame in Italian eyes.[5]

After the liberation of all Italy, in May, 1945, the Action and Socialist parties had rejoined the other four "Leftish parties"[6] which had made up the Committee of National Liberation which, in turn, had been the guiding body for the Par-

4 Others were National Democratic Union, 41; Common Man, 30; Republicans, 23; National Freedom Bloc, 16; Action, 7; and eight other parties with a total of 13. Cf. *Facts on File,* VI (June 18, 1946), 206.

5 See M. F. Neufeld, "The Failure of AMG in Italy," *Public Administration Review,* VI (Spring, 1946), 137–48.

6 W. S. Churchill, *Closing the Ring* (Boston, 1951), p. 188.

tisans. Parri had been the leader of the Partisans in the North, where they had been most effective because they were the best organized. Despite Allied use of the Italian Partisans in some areas, and despite their own initiative and success in some major operations, such as the capture of Milan prior to the arrival of Allied troops, they were not generally treated as "allies" by the Allies. Various Allied leaders paid verbal tribute to some of their more spectacular efforts and AFHQ released occasional bulletins praising them,[7] but they were treated tactlessly and belittlingly. Their very success and apparent organization were against them, for the Allied troop commanders, knowing little of Italian politics or feelings, feared these armed guerrillas who roamed their native areas at will on all sides of the Fifth U.S. and Eighth British Armies. One can understand the reluctance of Allied military men in having "foreign" armed men whose politics, organization, and intentions they did not know operating freely in the Allied military zones. But political leaders and AMG personnel could have found and insisted on a better solution than the one adopted which was immediate and wholesale disarmament of the Partisans in all areas.[8] The situation was of course complicated by the mushrooming of thousands of eleventh-hour "Partisans" who opportunistically donned a red neckerchief or arm band and, arming themselves, joined the victory band wagon to be "in at the kill." Nevertheless, it was a severe blow to the pride of those who had done good work—and in some cases magnificent work—in helping the Allies rid Italy of the Germans. Parri's popularity and hold on his own people were seriously weakened by this Allied psychological blunder.

[7] Cf. Churchill, pp. 187–88; M. W. Clark, *Calculated Risk* (New York, 1950), p. 381; *Facts on File*, V (March 16, 1945), 230; HQ, Fifteenth Army Group, *Finito: The Po Valley Campaign, 1945* (Milan, 1945), pp. 55, 59; and *A Brief History of II Corps* (Milan, n.d. [*ca.* 1946]), p. 64.

[8] It would seem that a solution which would have ended the intolerable confusion but which would also have included the elements of gratitude toward, and preservation of the pride of, the Partisans would have been to attach them, as military units, to the regular Italian troops then operating with the Fifth and Eighth Armies. The latter could have screened out the undesirables and then could have gradually demobilized and returned home others with some military fanfare and recognition, such as discharge certificates and campaign ribbons and decorations where indicated.

THE PEACE TREATY AND ITALIAN REACTION

The peace treaty itself was of considerable political importance to Italians and its terms were generally resented as too harsh, especially under the circumstances obtaining after 1943. Italians felt that their efforts in behalf of the Allies had been for nought. This treaty appeared to put them in the class of a defeated enemy rather than a cobelligerent.[9] It is interesting that in a country where the normally poor communications had not yet recovered from extensive war damage, more than three-quarters of the men and more than half the women knew something about the terms of the treaty. Almost half the men and one-quarter of the women had *full knowledge*[10] of its terms. In October, 1946, the Italians were asked, "Which of the Big Four disappointed you most with regard to its attitude toward Italy at the Peace Conference?" The Soviet Union led, in that 32.2 per cent of the Italians voted for it as the most disappointing. This is probably the result of the U.S.S.R.'s support of loss of land to Yugoslavia and of its insistence on reparations from Italy. France got 21.1 per cent of the votes, probably also due to its taking certain Italian land and perhaps because of the close, and often unfriendly, relations between the two countries. What is surprising is the discrepancy in the vote accorded the two major Allies who fought in and temporarily controlled Italy: the United Kingdom was voted "most disappointing" by 27.2 per cent of the Italians, while the United States "most disappointed" only 6.2 per cent of them.[11]

A FUND OF AMERICAN POPULARITY

One may hazard some guesses as to the cause of this marked difference in Italian feeling toward the two great Anglo-Saxon partners. The British, for various reasons, were the "senior" partner in Italy, particularly in the early days of the war, and apparently bore the major portion of the blame for what may be called political shortsightedness. Also it is probable that,

9 See *New York Times*, July 27, 1946; and *Facts on File*, VI (December 6, 1946), 320.

10 DOXA Bulletin No. 1, 1947.

11 DOXA Bulletin No. 2, 1947.

being European and closer to Italian history—ancient and recent—their faults were more apparent and their intentions more suspect to the Italian people than were those of the distant and somewhat idealized America. Perhaps a minor factor was the number of American soldiers of Italian descent with whom the natives had come into contact, as well as the free and easy ways of the American soldiers with their money. American soldiers abroad are more apt to make fools of themselves, and often fail to gain the respect the British do, but they also are more apt to make friends, since they lack the reserve which is characteristic of the English, at least. An opinion poll taken in March, 1947, shows the difference in Italian feeling toward soldiers. The question asked was, "Of all the Allied soldiers whom you have seen in Italy, which made the best impression on you?" Sixty-two per cent said American. Only 14 per cent voted for the British Commonwealth troops, including English, Scots, Irish, New Zealanders, Palestinians, and South Africans.[12]

Almost certainly a major factor of America's standing in Italian eyes was the aid given. Aid under military auspices has been mentioned earlier. From late in 1943 until August, 1945, the military aid to prevent "disease and unrest" was the responsibility of the Allied Control Commission and amounted to approximately $500,000,000. Because of its greater wealth, much of this burden fell on the United States, and the Italians were not unaware of that fact. From September to December, 1945, the United States took over the entire burden of an additional $100,000,000, which it administered through the Foreign Economic Administration. Another $120,000,000 in foreign currency came to Italy as a result of an arrangement whereby U.S. troops were paid in lire[13] while the Italian government received the dollar equivalent. While UNRRA has been much maligned and undoubtedly made many mistakes and suffered from some outright mismanagement, it saved

12 DOXA Bulletin No. 2, 1947 (poll taken March, 1947). Sixteen per cent had no opinion and 8 per cent voted for "others," who were in Italy in much smaller numbers and included Brazilians, French, Poles, French Arabs from North Africa, etc.

13 The use of the word *lire* refers to the occupation scrip in which all troops were paid. It was "legal tender" for the Italians also.

many lives and may even have saved Italy. It was an international agency and its officials were of various nationalities, most of them devoted to their common, humanitarian purpose. There is no evidence that any of the contributing nations consciously attempted to gain prestige from its operation. Nevertheless, the fact that many of its officials and the bulk of its funds were American could hardly have escaped Italian notice. UNRRA brought $50,000,000 worth of goods into Italy in 1945 and about ten times as much in 1946.[14]

THE GROWING MENACE OF THE EXTREME PARTIES

The year 1946 saw a continuous political writhing in Italy. The De Gasperi coalition Cabinet, formed late in 1945, contained representatives of all major parties. But throughout the year, it became increasingly apparent that the Communists and certain elements of the Socialists were participating principally to obstruct rather than assist any positive moves forward. Many Italians were jobless, many were near starvation. Unrest and discontent and governmental incapacity were widespread. Strikes were frequent, some honestly protesting conditions, others designed by the extreme Left as political cudgels with which to flay the government. There were growing indications that a new Fascist-type movement was afoot. Riots, mob scenes, and smaller disturbances were frequent. In municipal elections in November in Rome, Turin, Genoa, Florence, and Naples the extreme Left won. In Palermo the extreme Right, neo-Fascists, won. The outside world became worried over Italy's future. It should be recalled that this was a period of postwar unrest everywhere. Greece was in the throes of a civil war, Russia and her allies "haggled over the larger part of the vacuum Hitler left."[15] In this setting Italy was rapidly forging to the front as a major problem.

14 See DOXA Bulletin No. 2, 1947, p. 3, and *Facts on File*, VI (October 27, 1946), 315. Cf. N. Standen, "Italian Backdrop," *Harper's Magazine*, CXCVI (February, 1948), 150–59, especially p. 159 for typical condemnation of UNRRA. It is true that much of its relief found its way to the black market, but much did not and all of it helped feed the people. Cf. D. L. Cohn, "You Can't Eat Democracy," *Atlantic*, CLXXVII (June, 1946), 59–62, which indicates the U.S. contribution to UNRRA.

15 H. K. Smith, p. 16. The real Cold War started later, but the seeds were surely germinating in 1946.

AMG finally withdrew from all of Italy except the provinces of Venezio and Udine in December, 1945, relinquishing complete control of the rest of Italy to native civilian authorities. Those provinces were retained under military control because of the unsettled Trieste and Yugoslav border questions. Allied censorship of the Italian press had been removed in June, and in July Italian prisoners of war still in North Africa were permitted to return home and regain their former military positions. In August, the Allies had permitted Italy to resume complete control of its foreign trade. Thus by the end of 1945 Italy was virtually sovereign once again through a series of magnanimous Allied gestures.

GROWING AMERICAN CONCERN AND AID

Throughout 1946, the United States continued to play a helpful and friendly role toward Italy. Assistant Secretary of State James C. Dunn was appointed Ambassador in July. In September the U.S. Foreign Liquidation Commission sold $565,-000,000 worth of surplus military (but non-combat) supplies to Italy for $160,000,000. These were largely trucks, desperately needed to help restore the distribution system. Fifty million dollars in cash was given to Italy to reimburse it for lire given the United States during the war for purchases in Italy. And in December it was announced that De Gasperi would make an official visit to the United States. An American Air Force plane transported him, and much publicity and show of friendship attended the visit.

The trip was crowned with a certain amount of success both for De Gasperi personally as a political leader and for closer ties between Italy and the United States. The Export-Import Bank promised a $100,000,000 loan; the United States diverted six wheat-laden ships en route to the American forces in Germany to Italian ports to help stave off a threatened cut in the bread ration; and, perhaps most important, the United States forgave Italy its entire debt for food and relief supplies furnished prior to UNRRA. But matters were far from stable. In February, 100,000 peasants marched on Rome demanding bread, work, and land, and in the following months of the spring workers and peasants rioted or demonstrated in various cities all over Italy demanding work, food, relief, tax relief, etc. There was growing evidence that certain ex-Partisans, probably Communists, had cached arms at the end of the war instead of turning them in as the Allies had ordered.[16] Foreign Minister Carlo Sforza was beaten by an unemployed mob in Rome; a machine-gun massacre of parading agricultural workers in Palermo and the assassination of an anti-Communist editor in Milan were indicative of the extremes of violence prevalent in the spring of 1947.

AMERICAN FAILURE TO EXPLAIN THE TRUMAN DOCTRINE

On May 12, President Truman made his famous "Truman Doctrine" speech. Without going into the pros and cons of the Greek situation, it will suffice to say that the Italian Left, perhaps partly on Moscow orders, but certainly at least partially because of its own spontaneous ideas, was sympathetic to the antigovernment faction in Greece and strongly opposed to the British line, now taken up by the United States. If we had a good case, and most of us believed we did, this would certainly have been a propitious moment to explain our ideas and our actions to the world, especially that portion of the world torn by many of the same problems besetting Greece. Yet at this juncture of unfolding events, the House of Representatives voted to cut Voice of America funds by $66,000,000. Some of this was later restored, but the program suffered a serious setback which was not rectified until 1948.[17]

THE EXTREME LEFT IS OUT OF THE GOVERNMENT

In mid-May the government fell and, finally at the end of that month a new one was formed by De Gasperi which, for the first time, excluded the extreme Left. Immediately, in various cities there were demonstrations at this "purge." Those elements which had done all in their power to make the government in-

[16] See n. 8 above. Aside from the politico-psychological error of ordering the "in-place" disarmament of the Partisans, it would appear that this later danger of hidden arms caches might have been averted or greatly minimized had the Partisans reported, armed, to Italian army units. They would then have been disarmed upon discharge without rancor or loss of pride.

[17] For a graphic and compelling idea of what happened to VOA in the spring of 1947, see H. Block, *The Herblock Book* (Boston, 1952), p. 73, the 1947 cartoon entitled "The Gasp You Just Heard Was the Voice of America."

effective while they were part of it now accused De Gasperi of treachery to Italy and of being a puppet of the Vatican and the United States. Political strikes were now authorized by the Communist-led General Confederation of Labor and the Communist press called for De Gasperi's ousting by democratic means. On June 14 the Constituent Assembly extended its own life to December 31, 1947, thereby postponing the general elections scheduled for October, 1947, to the spring of 1948. The extreme Left accused the government of engineering this delay to avoid being beaten at the polls. While there may well have been some truth in this claim, it does not relieve the accusers of their own continuous obstruction and delay in the Assembly.

America Intervenes Openly

Violence and strikes continued in scattered areas all over Italy for the remainder of the year, and despite congressional blindness in regard to propaganda, President Truman's administration had become increasingly aware that Italy was in imminent danger of going Communist. On June 5 Secretary Marshall first proposed the plan which bore his name, and new hope rose in all those not already distrustful of, or dedicated to oppose, the United States. In August the United States canceled one billion dollars of war claims against Italy and released $60,000,000 of Italian assets "frozen" in the United States since 1941. Twenty-eight surplus merchant ships were given to Italy in September to help make up for her war losses. And in the same month America redeemed ten to fifteen million dollars worth of scrip given in payment to Italian prisoners of war while they worked in captivity. The United States also agreed to the negotiation of a new pact of friendship. Late in September the President asked Congress to take emergency measures for stopgap aid to see Italy through the winter until the Marshall Plan could be worked out.

CONGRESS BECOMES AWARE OF A NEED FOR PROPAGANDA

Fortunately for Italy as well as for the free world in general, a large number of congressional committees and subcommittees had spent the summer months of 1947 in official investigatory visits to various countries, including Italy. A comparison of their attitudes before these visits, as evidenced for example in the VOA slash, and their attitudes afterward attest to the fact that they saw much and were alarmed by what they saw. The Herter subcommittee of the House Foreign Affairs Committee and the Smith-Mundt Congressional Group investigating U.S. information services were two of the most aroused and most influential in restoring to the United States the ability to exercise its role as leader of the free world. The latter group's summary of the situation in Italy in September and October, 1947, is revealing of the scope of Congress' enlightenment.[18]

> Italy is presently going through a social, political and economic crisis. Although the majority of the population wishes to continue to live within the general pattern of a social and economic democracy there is today a growing doubt whether Italy can solve her problems within the framework of a representative democracy. The country is under great pressure from within and without to veer to the left and adopt a totalitarian-collective national organization. The extent of this pressure can be judged by the fact that there exists in Italy one of the largest Communist parties outside of the U.S.S.R., with a membership of nearly 2½ million. At the same time, dissatisfaction with the economic situation, with the Communist-inspired disorders and with the terms of the dictated peace treaty is leading many patriotic Italians to believe that a solution to the present difficulties lies only in the formation of an ultra nationalist, conservative-totalitarian organization. . . .[19]

These further gestures of good will took place in the last months of 1947. The United States, in early October, followed a few days later by Britain, waived its treaty claims to Italian warships. On November 6, Ambassador Dunn went to Venice to welcome the one-hundredth American relief ship to arrive in Italy in the preceding nine weeks.[20] Secretary Marshall

18 U.S. Congress, *Report of the Committee on Foreign Relations,* Appendix, 80 Cong., 2 sess., Report No. 855, Part 2 (Washington, 1948), pp. 199–210—hereafter referred to as the Smith-Mundt *Report.*

19 P. 203.

20 Mr. Dunn here started his habit of detailing, in his speech, just how much aid America had furnished. These 100 ships, bearing post-UNRRA free aid to Italy had carried 238,000 tons of wheat, 7,300 tons of other food, 615,000 tons of coal, 6,000 grams of streptomycin, 40 billion units of penicillin. *Facts on File,* VII (November 6, 1947), 367.

recommended that Congress approve $227,000,000 for immediate stopgap aid to Italy. And on December 19, Congress appropriated the stopgap funds. On December 29, the first "Friendship Train" was turned over by Ambassador Dunn to Vice-Premier Pacciardi in Naples. In accordance with the treaty, the last American troops, less than 2,000, sailed from Italy in mid-December. The once mighty U.S. forces of over half a million had all left now and Italy was free of the U.S. military, but on that day President Truman issued a warning that they would return if Italy's peace and freedom should be threatened. American warships, now in strength in the Mediterranean, had for some time been making calls at all Italian ports capable of handling them. Though correctly done in accord with international protocol, there can be little doubt that the frequent presence of these ships of the mighty United States of America in all of the major ports was designed, at least partially, as a psychological move (which the U.S.S.R. protested) to impress the more extremist elements of the population and to reassure the great inarticulate masss of the Italians that American strength was not distant.

THE VOA AND ITS ITALIAN AUDIENCE

The Voice of America had been cut in funds and therefore in activity, but it was not dead and some programs continued for Italian listeners throughout 1947.[21] There were four shortwave programs (on different call signs) between 8 and 9 P.M. and four between 10:30 and 11:30 P.M. daily, beamed in Italian. The relayed programs suffered most from the budget cut; four weekly, fifteen-minute programs had to be dropped. The daily five-minute program remaining consisted of American editorial opinion and reached the Italian public on their

[21] Smith-Mundt *Report,* p. 205, gives the following comparative data for U.S., U.K., and U.S.S.R. radio activity in Italy for mid-1947.

	Direct Short Wave	Relayed Rebroadcast	Recordings Transcribed
U.S.	8–15 min. daily	1–5 min. daily news 1–30 min. weekly	2–30 min. weekly and 100 hr. monthly from USIS records
U.K.	4–8½ hr. weekly	1–5 min. daily 2–30 min. weekly	some
U.S.S.R.	3 hr. daily	1–30 min. weekly	none

N.B.: U.S.S.R. has also satellite stations from Yugoslavia and Czechoslovakia beamed to Italy, plus the effect of native Communist party radio appearances.

own wave lengths at 2:20 P.M. The University of the Air program, also on long wave, was a weekly feature with Italian and British participation. Transcribed rebroadcasts included the weekly musical programs "Musical America" and "Hit Parade." In addition, the United States Information Service (USIS) provided Italian stations with about 100 hours monthly of American musical recordings.

A DOXA poll in 1949[22] found that about 28 per cent of all Italians did not listen to the radio. The percentage may have been somewhat higher in 1947, although comparable figures are not available. This seems a likely assumption because of the greater proximity of the disruption of war. About 35 per cent listened to the radio regularly, the remaining 37 per cent listened irregularly.[23] About 25 per cent used the radio as their only source of news, as they read no papers.

USIS SERVES THE PRESS

Because most publicity in America has attached to VOA, many Americans think that was our only, or at any rate our principal, propaganda medium. It could perhaps be most effective, since radio reaches out over space and is available for all who will, or can, to listen. But the other U.S. propaganda media active in Italy in 1947 were numerous. About 1,000 Italian editors received the daily radio bulletin, in Italian, put out free by USIS and covering U.S. news, especially U.S. foreign relations news. Usually this bulletin was accompanied by a special feature, which might be a biographical sketch or picture of an American leader, or a more detailed "background" article on some U.S. speech, policy, or activity of current interest to Italians. USIS also furnished Italian editors a monthly index of the official statements which had appeared in the daily bulletins and about

22 DOXA Bulletin Nos. 9, 10, June, 1949. The discrepancy of 2 per cent between the 25 per cent who read no papers in 1949 and the 23 per cent who read no papers in 1947 (see n. 39 below) may be a polling error or it may reflect the greater number of radios and radio stations in operation in 1949.

23 In late 1948 there were about 2,500,000 radio sets in Italy, or one per eighteen people. U.S. Department of State, *The World Audience for America's Story* (Washington, 1949), p. 92 (booklet based on a report by the President's Advisory Commission on Information). The number of sets was probably a bit lower in 1947. Unlike radios in the United States, a rather high proportion in Italy (1,500,000) are equipped for short-wave reception. See *ibid.*, p. 109.

sixty-five untranslated, scientific features monthly.[24] Commercial outlets for American news available to Italians included the *Rome Daily American, Time, Life,* and the Paris edition of the *Herald Tribune.* Italian editors could subscribe to AP, UP, INS, ONA for news items as well as photos and features.

ITALIAN NEWSPAPERS AND READERS

DOXA has produced some valuable data on Italian newspaper and periodical reading habits. In 1947 over 77 per cent of Italians over 18 years of age read newspapers and something under 23 per cent read no paper. Some of the readers read only a weekly paper and some saw a paper even less often.[25]

Further information on Italian reading habits was obtained in a poll in December.[26] It was determined that 6 per cent of adult Italians read periodicals but never read newspapers. That leaves only 17 per cent of the population who read nothing which would be apt to inform them of domestic or international events. This is rather a remarkable readership in a country as poor as Italy and with an illiteracy rate of about 16 per cent.[27] But it must be remembered that much of this reading is irregular and even then often what is read may be largely misleading and more emotional than factual. Nevertheless, the

24 All figures on USIS activities come from the Smith-Mundt *Report,* pp. 204–10.

25 DOXA Bulletin No. 15, September, 1947, gives the following percentages of readers, both regular and irregular: North, 85.1 per cent; Central, 79.7 per cent; South, 63.5 per cent; islands, 64.7 per cent; employers, 92.7 per cent; farmers, 78.9 per cent; farm laborers, 63.5 per cent; craftsmen, 87.7 per cent; workers, 81.1 per cent; white-collar workers, 89.2 per cent; executives, 88.5 per cent; housewives, 69.2 per cent.
NOTE: DOXA included under "farmers" both the large landowner and the small, landowning peasant; under "workers" skilled and unskilled non-agricultural employees; under "craftsmen" the self-employed artisan and those employed in very small shops; under "executives" managers and independent professionals, such as doctors; under "housewives" students and retired pensioners.

26 DOXA Bulletin Nos. 21, 22, December, 1947. The reading habits of an "audience" are important data for cruciality and susceptibility studies.

27 Recent statistics are unavailable. The figure given is an estimate taking into account the known rate of literacy increase for years before 1931, but making an arbitrary allowance for the disruption which the war years caused in the Italian school system. A rate of 5 per cent was thus arrived at for the years 1931 to 1947.

Italian population is fairly well aware of current events, and, judging from an amazing number of "Don't Know" answers found in most American polls, perhaps as well informed as Americans. As will be shown later, Senator Capehart was very mistaken when he said that not one-tenth of 1 per cent of the Italians would know anything about the American aid program.[28]

USIS AND THE CINEMA

USIS was as active in the field of films in 1947 as budget limitations allowed. Five USIS offices in Italy (Rome, Milan, Florence, Naples, and Palermo) each stocked 100 titles of educational and documentary films. The Rome office held an additional 30 medical films. Only about 40 per cent of all these films were in Italian, however. Even so, about 100,000 Italians witnessed USIS films each month in 1947.[29] Many of the showings were in provincial areas and dealt with a variety of subjects designed both to give the Italian some idea of America and Americans and to help Italians improve their agricultural and industrial techniques. USIS showed a selected group of Italian industrialists 72 industrial films and, based on their critiques, made available for use in factories, workers' clubs, and technical schools those believed most useful and effective. Because most organizations such as these and most small provincial communities lacked facilities, USIS films were usually shown on mobile U.S. equipment and often outdoors when weather conditions permitted. Arrangements were made with the local authorities, and news of the coming cinema usually traveled by word of mouth. A sound truck and announcements and music just before the showing acted as reminders to bring out an audience. This was not difficult to do in rural areas where any film was a great novelty and welcome diversion in an otherwise dreary and isolated existence.

Six hundred American commercial films were registered for import to Italy in 1946 compared to 100 British and 30 Russian. This was the only area in which American "information" outweighed that of the two other Great Powers. More were

28 *New York Times,* March 16, 1948.

29 Smith-Mundt *Report,* p. 205.

imported in 1947 but films travel more slowly and become "obsolete" much less rapidly in Italy than in America, so that it is likely that 1946 imports were still being shown extensively in 1947. The thinking Americans abroad are wont to decry the selection of American films shown to foreigners.[30] It is certainly true that some strangely distorted impressions of America can be gained from our movies—impressions which do little to build respect for us or solidarity in the free world. Nevertheless Italians, like most Europeans, enjoy American films. Considering the general level of poverty and the other difficulties involved, a surprising number of Italians went to the cinema remarkably often in 1947. Over one-third of all Italians between 26 and 55 years of age were once-a-week movie-goers. But this average was made possible by city dwellers because in towns under 5,000 over 35 per cent of the people claimed never to have gone to a movie.[31] It was to these, of course, that USIS mobile film units were apt to be most appealing and most meaningful.

LIBRARIES AND EXCHANGE OF PERSONS

The radio, press, and movies constituted what might be called the "active" aspects of U.S. propaganda in Italy. There were also more "passive" activities. The five USIS offices each maintained libraries of some 2,500 books and 200 to 300 periodicals. Four "reading rooms" were in operation, in Genoa, Turin, Bologna, and Bari. These various means of Italian-American contact were reported to reach about 10,000 to 12,000 Italians a month.[32] Commercial books from American sources were scarce and expensive in Italy in 1947, although about 200 titles had been translated in Italian since the war. American publishers were not interested in lire and the Italian government was reluctant to permit its dollar reserves to be tapped for the importation of American books.

Another "passive" propaganda medium was the exchange of persons. The program in Italy in 1947, from a national view-

[30] Cf. W. Wagner, "OWI and Motion Pictures," *Public Opinion Quarterly*, VII (Spring, 1943), 100–110.

[31] DOXA Bulletin No. 18, November, 1947.

[32] Smith-Mundt *Report*, p. 206.

point, was negligible. Only 80 Italian students came to the United States on a scholarship or fellowship basis and only about 100 Americans were studying in Italy on scholarships or the GI Bill. There were probably a handful in each country on their own funds, but no records are available. There were also American tourists in Italy, but these are not always a good "advertisement" for the United States. The most notable Italian visitor to the United States in 1947 was the Prime Minister. The success of his ten days, as measured in concrete gains for Italy, has been mentioned.[33]

A little-known activity of USIS was its assistance to organizations and activities in furnishing information on America. Seven Italo-American associations (in Rome, Genoa, Milan, Turin, Venice, Florence, and Naples) served as means of exchanging books and periodicals, traveling exhibits, and photo exhibits. The USIS cooperated with these on festivals, fairs, conventions, concerts, etc. Over 13,000 American publications were furnished to firms and societies; USIS office library and reading room windows and interiors themselves were points of attraction for the Italian visitor and passerby, as they always contained enlarged photos and other material showing some aspect of America or American life and customs.

ITALIAN ATTITUDES TOWARD THE UNITED STATES

What was the effect of all this activity? Was the Italian aware of American aid? What did he think of the United States? Fortunately there are some valuable opinion polls on these points during 1947. To the question, "Which of the Big Four do you consider best disposed toward Italy?" 75.8 per cent of the Italians responded, "The United States." Only 8.7 per cent thought it should be Russia and even fewer voted for Britain or France.[34] Even among industrial workers and hired farm laborers, the most "Red" elements in Italy, the United

33 Ambassador Alberto Tarchiani wrote a book, *America-Italia: Le dieci giornate di De Gasperi negli Stati Uniti* (Milan, 1947), about this trip and its success. While not "American propaganda," it certainly did America no harm in the minds of its Italian readers. Cf. also Banco di Roma, *Review of the Economic Conditions of Italy*, I (1947) 18–20, for a glowing report of De Gasperi's visit to the United States.

34 DOXA Bulletin No. 2, February, 1947.

States received a clear majority of 64.3 per cent and 61.7 per cent, respectively. Surprisingly enough, 31.7 per cent of the interviewees who were avowed Communists thought that America was best disposed toward Italy, as against 50.3 per cent who believed the U.S.S.R. was. Another indication of America's high standing among the Italians was the fact that 17.9 per cent of them thought an American should be governor of Trieste. Switzerland, with 19.8 per cent, was the only nationality preferred over the United States, and Russia received only 3.2 per cent of the vote.[35]

Perhaps indicative of a feeling that Italy was a pawn in the world power struggle was the fact that only 5 per cent of the Italians felt that no nation sought world dominion. A third of

TABLE 3

	PCI*	PSI	PSLI	PDC	PLI	UQ	PNM	IND
U.S.S.R.	8.1%	9.0%	25.9%	42.7%	42.8%	34.2%	29.1%	30.1%
U.S.A.	34.1	16.1	3.4	3.5	0.8	5.0	9.7	5.8
U.S. and U.S.S.R.	17.7	32.2	41.4	28.0	33.8	33.3	45.1	34.8
DK† and other	40.1	42.7	29.3	25.8	22.6	27.5	16.1	29.3

* The designations signify: PCI, Communist; PSI, Socialist; PSLI, Labor Socialist; PDC, Christian Democrats; PLI, Liberal; UQ, Common Man; PNM, Monarchists; IND, Independent. These abbreviations will be used hereafter.

† DK signifies "Don't Know."

the nation felt that the United States and the U.S.S.R. both wanted to dominate the world; 31 per cent thought only the U.S.S.R. did and 17 per cent thought only the United States did. A revealing division of opinion along partisan lines appeared in the latter part of 1947. The question was, "What nation or nations do you believe want to dominate the world?"[36] (See Table 3.)

American readers will at once be impressed by the fact that, except for the extreme Left (PCI and PSI), few Italians believed the United States alone sought world domination. But before allowing too smug a feeling to overcome reason, a long look should be given to the number of Italians of all parties who lumped the United States and the U.S.S.R. together in this class.

35 DOXA Bulletin No. 3, March, 1947. The question excluded Italians and Yugoslav citizens as possible governors.

36 DOXA Bulletin No. 16, September, 1947.

THE ECONOMIC SETTING FOR THE ELECTIONS

The stark economic desperation of the country should be borne in mind for this was a "backdrop" for all other considerations.

The free soup kitchens of Genoa run out of food long before the customers' pails are filled. Ragged children dive under cafe tables for butts which their unemployed parents make into "new" cigarettes . . . and sell for fifty cents a package. In Florence, fifteen-year-old prostitutes hang outside the Allied hotels begging soldiers to pay them with goods, and, as part of the deal, to feed the kid brothers and sisters clutching their skirts. The slums of Naples match those of Cairo and Bombay. From everywhere, men and women of all ages flock to the Swiss, French and Belgian recruiting centers to apply for work, but more than half are refused because of malnutrition and tuberculosis. Spaghetti is a major luxury; most Italians . . . exist on watery soup and corn meal.[37]

These were not isolated and unusual cases in a country where almost 60 per cent of the families earned less than $825 a year and 93.8 per cent received less than $2,070 a year.[38] Compared with prewar conditions, 54 per cent of Italian families were living on less food and 56.9 per cent were more poorly clothed. But some had profited from the war and postwar conditions; 14 per cent were better fed and better clothed. For those who had held their own in the matter of living standard, as well as for those who had been lowered in status, economic problems were ever present and the source of greatest concern. To the question of what were the most urgent problems faced by the government, over 80 per cent of the people replied with some aspect of economics, whether unemployment, wages, inflation, or prices.[39] Unable to help themselves, over one-quarter of the population would emigrate permanently if given the opportunity[40] and the bulk of them thought the government should increase rather than decrease

[37] Standen, p. 150 (1947 conditions).

[38] DOXA Bulletin Nos. 21, 22, December, 1949. (Poll taken last part of 1947 and first part of 1948.)

[39] DOXA Bulletin No. 9, June, 1947.

[40] DOXA Bulletin No. 4, April, 1947.

its activities in the economy of the nation.[41] This was the setting for the intensified political campaign which began along with the year of 1948.

THE POLITICAL SETTING FOR THE ELECTIONS

Under the new constitution, which became effective on January 1, 1948, it was easy to form a new political party and to have its name entered on the ballot. Over three hundred, most of them purely local, were in legal existence by election time in April. From "Left" to "Right," the major parties were: the Communists (PCI);[42] the Socialists (PSI); the Labor Socialists (PSLI); the Republicans (PRI); the Christian Democrats (PDC); the Liberals (PLI); the Common Man (UQ, for *Uomo qualunque*); the Monarchists (PNM); the Italian Social Movement (MSI). The Communist was the party "to beat" in the 1948 elections and some description of that party appears appropriate. The Communist party in Italy was the largest in the world outside the Iron Curtain. It emerged from the war, and German occupation and Allied occupation periods as the best-organized and strongest political force in the nation. In some ways it was probably stronger than the church.[43] It had remained an underground apparatus under Mussolini and various purges, both internal and external, and had strengthened its core of leadership so that only the staunchest and most dedicated remained. As fascism waned in Italy, this leadership began to exercise increasing influence among all Italians who, regardless of political affiliation, were sick of tyranny, whether Italian or German. The Communists, drawing on the new spirit of *Risorgimento* awake in the land, offered action and excitement in the process of assisting in Italy's liberation. The hard core of disciplined leaders were sufficiently wise and dis-

[41] DOXA Bulletin No. 9, June, 1947.

[42] Italian initials will be used throughout to avoid confusion. The Italian name of this party is *Partito Comunista Italiano.*

[43] Cf. L. Olschki, *The Genius of Italy* (New York, 1949), p. 460, where it is contended that the Church was stronger; A. Garosci, "The Italian Communist Party," *Communism in Western Europe,* ed. M. Einaudi (Ithaca, N.Y., 1951), pp. 176–77, wherein some of the organic strength of PCI is indicated; A. O. McCormick in the *New York Times,* March 29, 1948, "All talk to the contrary, the Church is not organized for a political campaign"; and M. Salvadori, *The Rise of Modern Communism* (New York, 1952), pp. 61, 106–7.

ciplined to seize the leadership of many of the Committees of National Liberation. They also took over the property, files, and equipment of the Corporative Labor organizations of fascism as that regime crumbled. The time and circumstances were ripe for spoils for the politically bold and the organizationally strong; the Communists effectively seized the opportunity.[44]

In a sense liberation was bad for the Italian Communist party, for no sooner had it arrived than the party had to show its true colors, which in many ways proved to be more Communist than Italian. The happy period of collaboration and cooperaiton of *all* Italians was over, as was the wartime cooperation of Russia and the West. Now came the time for internal bickering and political maneuver paralleling the wrangling of the Great Powers. The automatic and often stupid opposition of the PCI to all constructive action by the successive governments had lost them many of their wartime adherents. Economics rather than ideology had kept the PCI in a strong position. There is fairly general agreement among students[45] of the subject that unemployment, poverty, and allied ills formed the motivation behind the membership of many if not most Italian Communists. The PCI membership was about 50 per cent industrial workers and about 33 per cent peasants and hired farm laborers, all suffering from unemployment, under employment, and related lacks of security. Only about 15 per cent was made up of middle-class people and intellectuals,[46] who, because of education and economic status, may be assumed to have been members for ideological rather than economic reasons. This is not to say that, given better living conditions, the rest of the membership would have fallen away. The real hard core, and it was fairly large, existed in the in-

44 For an excellent treatment of the attractions of the CPI see Gabriel A. Almond, *The Appeals of Communism* (Princeton, 1954); and Hadley Cantril, *The Politics of Despair* (New York, 1958).

45 Cf. V. Ivella, "Party Rule in the Democratic State," *Foreign Affairs,* XXVIII (October, 1949), 75–83; F. Oppenheim, "Prospects for Italian Democracy," *Public Opinion Quarterly,* XI (Winter, 1947/48), 572–80; and N. Jucker, "Italy: East or West?" *Political Quarterly,* XXII (April–June, 1951), 175–85.

46 Einaudi, p. 217. Cf. P. L. Fegiz, "Italian Public Opinion," *Public Opinion Quarterly,* XI (Spring, 1947), 92–96.

dustrial element, and certainly some of the agricultural membership contained some "believers" in communism.[47]

In the 1948 campaign the PCI suffered because of the Communist world's attitude toward the Trieste question. The Italian leader, Palmiro Togliatti, tried hard to convince Italians that an Italian Communist government would have no difficulty negotiating a satisfactory settlement with a Communist Yugolsalvia,[48] but events tended to undermine the effectiveness of this appeal. On the other hand, its strength was enhanced by its control of the Italian Confederation of Labor (CGIL), which, though it had a membership of only about one million, through its organization, discipline, and reputation had a tremendous influence on all the Italian working class.[49] But members alone do not give strength to a political movement or party. The faith, sometimes almost religious in its intensity and irrationality, which provides the cohesion and the driving force of the PCI came largely from the intellectual and professional elements in the party. Salvadori says it was a "strange mixture" of "vision" and "practical instruction" of "idealism and the most sordid of practical politics."[50] Togliatti was the intellectual, capable of effective debate and thoughtful speeches, but inclined to "long-windedness." Luigi Longo, an effective Partisan leader, was generally conceded to be the No. 2 man in the party and it was his province to undertake forceful action, if such had been determined necessary. Both Togliatti and Longo were elected to the Cominform in October, 1947, thereby openly avowing the ties of PCI to Soviet-dominated world communism. The PCI won 4,342,722 votes in the 1946 elections.[51]

47 See Salvadori, p. 107, for a summary of the factors of Communist strength.

48 See P. Togliatti, *Il trattato di pace e l'Italia nel mondo* (Rome, 1947), pp. 20–21.

49 See Ivella, "Italy's Cabinet Strives To Avert Crash," *World Report,* June 24, 1947, p. 13.

50 M. Salvadori, "Italy," Foreign Policy Association, *Headline Series,* No. 87 (May–June, 1951), p. 31. For a discussion of some of the characteristics of the CPI leadership see Robert T. Holt, "Age as a Factor in the Recruitment of Communist Leadership," *American Political Science Review,* June, 1954, pp. 486–99.

51 K. M. Smogorzewski, "Italy after the Elections," *World Affairs,* II (July, 1948), 284–95.

By 1948 the Socialists (PSI) of the Left Wing variety led by the crowd-swaying orator Pietro Nenni were so closely allied to the Communists as to be barely distinguishable. This was the "rump" party after the departure of Saragat and his "Labor Socialists." The PSI was always Marxist, and after the war grew increasingly Leninist and even Stalinist. They were believers in the necessary unity of all the world's working classes and the impossibility of war between true Socialists. They saw the United States as the major, if not the sole, threat to peace. Nenni had said that Europe's workers must unite to defeat America's intention of fighting the U.S.S.R. to the last European.[52] For the 1948 election campaign, PSI formed with PCI the People's Democratic Front (FDP), in order to capitalize on the pooled vote of the adherents to each of the parties. In the 1946 elections PSI (which then contained that element which left with Saragat) received 4,744,000 votes, or about 2 per cent more than PCI had.[53]

The Labor Socialists (PSLI) of Giuseppe Saragat or, as they came to be called, the Social Democrats, were born of the split in the old Socialist party early in 1947. They too professed to be Marxists, but they were obviously first of all Italians, as they offered to enter the De Gasperi anti–extreme Left coalition government and did so early in December, 1947. After that, they usually were aligned with the Center parties in a pro-Western and pro-U.S. attitude. Their Marxist ideology is tempered, at least in its international sense, by their loyalty to Italy and their belief in cooperation and gradualism. Idealistically the PSLI has continued to hope for the eventual reunion of all Socialists, but not at the cost of U.S.S.R. domination. On the basis of the 5 to 6 division of deputies, it might be said that PSLI's share of the 1946 Socialist vote would have been about 2,155,000. With 53 deputies in early 1948, PSLI was Italy's fourth largest party.

The Republican party (PRI) was the holder of an ancient heritage of antimonarchism when it returned to life during the liberation period. Technically PRI lost its principal *raison*

52 P. Nenni, "The Responsibility of Left Socialism," *Monthly Review*, I (March, 1950), 345–48.

53 Smogorzewski.

d'être when the Republic was formed and the Monarchy abolished in 1946, but its democratic liberalism and the quality and vitality of its leadership had caused it to keep a certain popularity among Italians. In the outside world Sforza, who had for some time called himself an independent, was its best-known member when the party rejoined the coalition in December, 1947. PRI earned just under 1,000,000 votes in 1946, which made them sixth in size.

The "Center," according to DOXA, was made up of the center elements of the Christian Democrats (PDC). They were the largest single party in Italy and were the lineal descendants of Don Luigi Sturzo's *Popolari* of pre-Fascist days. The PDC was actually a loose coalition of moderate political views and the Christian (particularly Catholic) ethic. Einaudi called it "a political movement of complex ideology, one attempting to gain permanent roots by being the image of many cross-currents and of conflicting interests."[54] It contained many internal cleavages but, except for its Left Wing, tended to lean more to the Right than the Left. Its real binding force was its common belief in Christian religious and moral principles. Politically it was a party of order, tradition, gradualness, and slow evolution. Inherent were all the weakness and strength of the paradoxes common to Christian Democracy where "Christian" means "Catholic." It was "dogmatic in spirit and democratic in methods, conservative in temper and progressive in social action, catholic in conception but national in policy, intransigent in principles but cooperative in practice."[55] Its leader, De Gasperi, believed in working in close harmony with as many as possible of the small parties which fringed his. Only thus, he reasoned, could Christian Democracy form a strong and realistic link between the government and the masses. It was this effort to include all views, to consider all factions, and to alienate as few as possible which gave rise to the saying that the government's policy was "never do today

[54] M. Einaudi and F. Goguel, *Christian Democracy in Italy and France* (Notre Dame, Ind., 1952), p. 28.

[55] Olschki, p. 464. Cf. also C. Neikind, "The Cocktail Party That Shook Italy," *Reporter*, III (November 21, 1950), 33: "The Christian Democratic Party—a loose federation of factions with a wide range of economic ideas and interests."

what you may find an excuse for not doing tomorrow."[56] The Christian Democrats emerged from the 1946 elections far ahead of any other party, with over 8,000,000 votes.

The Liberal party (PLI) was misnamed for the twentieth century. It was the inheritor of the eighteenth- and nineteenth-century concept of laissez faire in economics and the least possible government interference in the "free" life of the populace. It was a conservative congregation made up largely of educated, respectable, and generally well-to-do upper-middle-class people who desired a return to "the good old days." It had fought with the various coalition governments on every program which might be called "liberal." It opposed any change in the ownership structure, even when monopolistic. It held sacred the right of property, even of known and prominent Fascists. It had broken with De Gasperi over the question of land reform. For the 1948 campaign it had joined to itself a portion of the Monarchists and formed the National Bloc. Benedetto Croce, probably one of Italy's best-known men and a famous philosopher, resigned as president of the party in December, 1947. The PLI was fourth in the 1946 race with something over 1,500,000 votes.

The Common Man party (UQ) was a rapidly formed hodgepodge of Rightist malcontents who gathered to themselves ex-Fascists and disgruntled Liberals and Monarchists and campaigned against the ineffectiveness of the government in 1946. They received well over 1,000,000 votes but were discredited in 1947 municipal elections for harboring ex-Fascists and were soon dissolved, with most voters returning to the more traditional parties, but some going to the new and almost frankly neo-Fascist *Movimento Sociale Italiano* (MSI). Throughout 1947 this group grew in size and in violence. They were responsible for several attacks on political persons and groups, especially of the extreme Left. The movement was a result of several factors—pauperism, psychological disorientation after the war and the fall of Mussolini, youth's desire for action, any action, which might ameliorate current conditions, and hate and fear of the PCI-PSI combine.

56 L. J. Wollemborg, "Italy: The Christian Democrats," *Reporter*, IV (June 26, 1951), 14–16.

The Monarchists (PNM) represented an anachronism in 1948. Even in 1946 when 10,000,000 Italians voted to retain the king, the PNM received only 636,000 votes. It was, even after losing a part of itself to the National Bloc, still a traditional force of the Right, important not for itself, but because it might be a link for merging the strength of PLI and MSI.

POLITICAL EVENTS PRIOR TO THE ELECTIONS

Between New Year's Day and the elections in April, there was no major Italian political event except the campaign itself. There were, however, several important events originating elsewhere which had a direct bearing on the campaign in Italy, either purposely so designed, like the Big Three announcement on Trieste, or accidentally relevant, like the Communist coup in Czechoslovakia. Nevertheless the campaign and its results constituted one of the most important events in the postwar world. It was "perhaps the most hectically fought political campaign that ever took place in Europe,"[57] and until the all-party truce of February 17, it was violent and promised to become increasingly so. Throughout January and early February a wave of political, Communist-inspired strikes and demonstrations swept the country. After the truce, peace was fairly well kept, but there was increasing evidence that the PCI-PSI might turn to violent means if they lost.[58] Early in February the PCI planned to bear part of the expense in uniforming the Partisans, and the government countered by issuing a decree outlawing all private associations of a military character. The Communists protested the decree saying the Garibaldi Brigades were no more military "than the Boy Scouts or the Catholic girls associations."[59] Yugoslavian sources were believed involved in smuggling in arms, and the Adriatic Coast

[57] H. K. Smith, p. 203.

[58] Cf. various reports of arms caches (usually those of Communist-Partisans who failed to turn them in after the war) in the *New York Times*, February 5, 15; April 3, 4, 6–11, 15, 18, 1948.

[59] *New York Times*, February 2, 6, 1948. The government's decree was in accordance with Article XVIII of the new constitution, but as yet no statute had been passed implementing this.

and border areas were carefully watched by the augmented internal security forces.

The last week in February saw the Communists seize control in Czechoslovakia, and the effect in Italy was electric. Only a week earlier, in municipal elections in Pescara, the Popular Front had won with a 10 per cent increase over its 1946 mark, while the Christian Democrats had barely held their own as a poor second. If anything was needed to spur on the anti-Communists, these two events supplied the impetus. On February 22, the Pope publicly called for support of the church against its atheistic foes. There was a trend noticed in a Labor election among Turin textile workers away from their traditional Left leanings,[60] and Anne O'Hare McCormick could report[61] that even some anticlericals had ceased condemning the church's political activity. The PDC used the Czech case as an example of what can happen when Communists gain a foothold in a democracy's governmental machinery and warned against certain "Front" activity in Italy as paralleling that in Czechoslovakia.

It soon became apparent to all Italians as well as to the world that the election campaign was boiling down to a miniature East-West struggle.[62] This is not to say that the Italians were not aware of the domestic struggle, only that all home issues were linked to and overlaid by the question of the United States versus communism.[63] In mid-March only about 4.3 per cent of the Italian people said they would not vote because of "complete disinterest in politics." About 70 per cent, on the other hand, believed the coming elections more important than those of 1946, and 16 per cent thought them equally important.[64] It is impossible to say that American prop-

60 *New York Times,* February 25, 1948. PDC candidates won 42 per cent and FDP 53 per cent, whereas at the previous election it had been 38 per cent and 60 per cent, respectively.

61 *New York Times,* March 3, 1948.

62 Cf. A. O. McCormick in the *New York Times,* March 10, 1948; and her "The Italian Political Scene," *World Today,* VII (July, 1951), 283–92.

63 See M. Ascoli, "Rules of Thumb for Foreign Aid," *Reporter,* III (August 29, 1950), 4–5.

64 DOXA Bulletin No. 8, April, 1948.

aganda and other operations having a psychological impact were *the* responsible factor for this upsurge of public interest, but it would be ridiculous to say that they were not influential.

THE CHURCH IN THE CAMPAIGN

Of course the Catholic Church and its agency, Catholic Action, reached many if not most Italians with appeals and with threats,[65] to vote against those who were a danger to religion and the church. The influence of the church in arousing the Italian masses cannot be minimized, nor can it be said that without American impetus that institution would not have undertaken as active a part in the campaign. It can be said, however, that the church appeared to be slow in commencing its activity,[66] and this may have been an indication that it did not fully realize the threat to itself until the issue had been widely publicized by others. It was not until mid-February that "Vatican circles" announced the church would "use its power and influence" to help the PDC. The church claimed this would not be a violation of the Concordat because it "is not political activity, since it is limited to defense of its own religious interests. . . ."[67] It is a fact that American anxiety, interest, and influence in Italian domestic politics long antedated this announcement. There is evidence too that the United States was not unaware of what the church could do in this campaign if it would. President Truman's sending Myron C. Taylor flying back to the Vatican from Spain on April 4 was certainly no mere coincidence. Nor was it coincidental that American Catholics, both clergy and laymen, were among the most active in the letter-writing campaign of which more will be said later. President Truman's November 17 speech on foreign aid had been couched in high moral lan-

[65] The *New York Times* issues from mid-February to election day almost daily carried reports of diocesan pronouncements threatening withholding of sacraments for those who vote "anti-Christ" or Communist, of parish priests' making the issue "For God or against God," of denoting abstinence from voting "a grave sin," etc., etc. All the Italian clergy was ordered to vote, even those orders of nuns who live in complete seclusion from the world. Togliatti denounced this activity as "spiritual terrorism."

[66] Cf. A. O. McCormick, *New York Times,* March 29, 1948.

[67] *New York Times,* February 13, 1948.

guage, but his speech to a joint session of Congress on March 17 was the first openly to emphasize the need for stopping communism. He stressed the threat to free institutions in Italy and asked "God's help" in American efforts to help Italy and other nations in similar trials. The Vatican openly acclaimed this speech.[68] A safe and completely fair conclusion is that America "connived at a political campaign by the Holy See . . ."[69] for ends each considered vital.

THE TREND AWAY FROM A COMMUNIST VICTORY

Despite American aid and propaganda (which have been detailed in chap. v), despite considerable recovery which had taken place since the war, it appeared in February and March that the forces of the extreme Left would be victorious.[70] And this would mean communism would soon control Italy, outflanking Greece and Turkey and the West's zones of Austria. A Communist-dominated state would thrust its spear deep across the Mediterranean and, as a close neighbor, greatly weaken France's chances of resisting a similar fate. About two weeks before the election, observers began to feel that the trend was changing, that the Left bloc would lose.[71] The church's contribution to this change has been very briefly summarized. It is most doubtful that it alone could have overcome the advantage of the early lead of the FDP and the intense and expensive campaign waged by the extreme Left.[72] The Communists had paper for newsprint, for posters with which they plastered most of the walls in Italy, for transportation; and they had manpower. Their control of CGIL gave them the support of 6,000,000 workers who looked to it for leadership. Moreover their efforts among the peasantry were

68 *New York Times,* March 19, 1948.

69 H. K. Smith, p. 207.

70 Cf. *ibid.,* p. 202; and *New York Times,* March 15, 1948.

71 *New York Times,* April 4, 1948.

72 A. Cortesi, *New York Times,* April 11, 1948, "The Communists, in these last few days before the election, as in the whole campaign, are spending a great deal more money than any other party—perhaps more than all the other parties put together."

both wide and ruthlessly persuasive.[73] What was the meaning of the change which took place in opinions?

As a matter of fact the "Front" really held its ground. In March, DOXA discovered that 19.7 per cent of the Italians thought the FDP should win in the interests of Italy whereas 38.4 per cent thought the Christian Democrats should.[74] Six days before the election a poll gave the two major rivals 27 per cent and 45 per cent, respectively.[75] In the March poll about 85 per cent said they intended to vote; whereas actually almost 93 per cent did vote. It was probably this group, largely in the "Don't Know" column in the March poll, which made it possible for PDC to gain a majority of seats in parliament. And it would appear probable that it was on these who were unsure, or fearful or disinterested, that American propaganda had most influence.[76] By a week before the elections only 5 per cent did not intend to vote.

Kinds of American Influence

The activities of the United States which all observers agree were influential in Italy in one way or another can be very roughly divided into four general groups: (1) direct, simple, and open propaganda to the Italian people; (2) economic aid to Italy which was used in such circumstances as to capitalize on its psychological effect; (3) actions or policies which, while perhaps directed toward another nation, were either designed or timed to have a symbolic effect in Italy; (4) actions or

[73] A favorite and frequently used expedient was for two party workers, one dressed in some sort of uniform, the other with pad and pencil being a "secretary," to call on a peasant family. The uniformed man would ask what the peasant and his family needed. The "secretary" would write it all down. Land was usually one of the items. A map would be produced and a plot of land decided on as desirable. It would then be marked with the peasant's name. All things desired were promised to the peasant if he would help elect the FDP.

[74] DOXA Bulletin Nos. 9, 10, April, 1948.

[75] DOXA Bulletin No. 11, April, 1948.

[76] Editorial, *Quaderni di Roma*, II (January–April, 1948), 141–42. Speaking particularly of the "letter campaign" waged by Italo-Americans, it says, ". . . could have some influence on the elections, but not enough to change firmly held opinions. And the proof of this fact is that the Communists maintained their position. The outcome was due to the arousing of that part of the electorate which formerly preferred to abstain [from voting]."

policies which were not designed or timed to have any symbolic effect in Italy but which did have such an effect "by accident."

OPEN PROPAGANDA AND PERSONAL APPEALS

Under the first group, of course, would fall the various USIS activities described in chapter v which continued in the period under discussion at as greatly a stepped-up tempo as budgets would allow. Most notable in this step-up was the increased programing of VOA on March 22. Daily programs were pushed up from two hours and five minutes to three hours. The series was planned give Italian hearers a better understanding of American aims and policies by giving them a fuller picture of American life. The programs included newscasts, which are always popular in Italy, especially when larded with news of assistance to Italy. Any new grant of aid or gesture of American friendship for Italy was, of course, played up, such as the progress being made on the Marshall Plan and the treaty of friendship, the return of gold seized by the Nazis (April 12) special grants to the Ministry of Fine Arts for the protection and restoration of damaged buildings in Italy (April 15), the ECA authorization for $8,000,000 worth of grains—second in one week (April 17), etc. The series included programs of music, which most Italians love and which frequently featured American musicians of Italian descent as well as other world-wide favorites. Also frequent talks by Americans of Italian birth or parentage who were leaders in business, labor, education, and agriculture were regular features.[77] Five broadcasts by Italo-American housewives were made. The object of this was multiple: to show the importance of the average woman in American life, to emphasize the success achieved by Italo-Americans (speaking on the radio indicates a certain prestige), to play on the common-heritage theme, to indicate

[77] Certain celebrities of the entertainment world, known to Italians, also made talks. Frank Sinatra was one; he not only has an Italian name, but toured Italy right after the war. Gary Cooper, voted "favorite actor" by Italians in a DOXA survey in the fall of 1948, was another voice used to persuade Italians to vote anti-Communist. He was a "high credibility source" as described in chapter iii, but using him was probably a result of intuition rather than research, as this DOXA poll was *after* the election campaign.

the concern of Americans of all walks of life for Italy and Italians. In March an all-American group gave a concert of American music in Rome, featuring several celebrated Italo-American musicians.

The VOA and government-sponsored activities were in addition to a whole rash of undertakings by various individuals and organizations, all attempting to sway Italians away from a FDP vote. Francesca Braggiotti Lodge, wife of U.S. Representative John Lodge, together with Common Cause, Inc., commenced a series of direct, daily broadcasts to Italy over WRUL in Boston.[78] These programs were heard in Italy at 8:15 P.M. on three different wave bands. Major General William Donovan, of OSS fame, broadcast over WRUL on March 22, saying that if the Front won it would mean the dismantling of Italian plants, which would then be shipped to Russia, along with workers.[79] He also emphasized that South Italy's problems could be solved only with outside aid and that such aid could come only from the United States. Acting Mayor Vincent Impellitteri broadcast over the same station, urging Italians to vote in such a way that "your beautiful country [will not] . . . become enslaved under the crushing heel of communism,"[80] and Arthur Bliss Lane, former Ambassador to Poland, broadcast a warning to Italians not to let Czechoslovakia's fate befall them. On April 12 WRUL increased its daily broadcasts to Italy from forty-five minutes to one and one-half hours. Part of the increased time was used to broadcast messages by refugees from behind the Iron Curtain, warning Italians of what life was like there. Part of this increase was made possible through the cooperation of Generoso Pope, wealthy publisher of *Il Progresso Italo-Americano,* New York Italian-language paper, and president of radio station WHOM.

78 *New York Times,* March 19, 1948, reported that this series had State Department approval. Common Cause was founded for this purpose by Mrs. N. W. Paine, who had founded Bundles for Britain during the war.

79 *New York Times,* March 23, 1948. Italians not only cherish their industrial plants, but they had very bitter and fresh memories of enforced labor in Germany.

80 *New York Times,* April 8, 1948. Regardless of what New Yorkers may think of Impellitteri's ability, he had prestige in Italy as an Italian immigrant "boy who made good" in the fabulous United States of America.

Mr. Pope planned a daily broadcast from WHOM in New York, to be relayed through WRUL. The series included talks and music. A special feature was the series of one-minute messages to relatives in Italy by Italo-Americans in the New York area. The *New York Times* reported on April 13 that hundreds of would-be message-senders had to be turned away from the studio on the opening day as time for only forty-five messages was available. On election eve this WHOM-WRUL network broadcast final appeals from Mayor William O'Dwyer, A. B. Lane, Attorney General Tom Clark, and V. Impelletteri asking for a "resounding victory for God and freedom."

VOA also made use of prominent names believed to have special prestige in Italy. Luigi Antonini, vice-president of ILGWU, not only was a well-known Italo-American, he was a prominent labor leader and was considered to bear some weight with Italian workers. William Green, president of the AF of L used VOA facilities to warn Italians of a repetition of the Czech coup and of the evils of slave labor. Others included Rhode Island's Governor John Pastore and U.S. Representative John Lodge. VOA also used the personal touch by having Italo-Americans make direct appeals to relatives and friends. Rome correspondent Arnaldo Cortesi warned, however, that too much faith must not be placed in the effectiveness of VOA. "It has a large following but principally among the intellectual classes who are mostly on the American side anyhow. It is seldom heard by the working people because few if any among them own radios."[81]

Organized American labor, in addition to the broadcasting mentioned above, entered actively into the campaign several times with direct propaganda efforts. David Dubinsky cabled to English leaders on the occasion of the International Socialist Conference, asking them why they had invited Nenni, since it merely helped Nenni to "promote a totalitarian dictatorship in Italy and to cripple the European Recovery Plan."[82] In reply to Front propaganda that American labor would not

81 *New York Times,* March 28, 1948.

82 *New York Times,* March 20, 1948. Nenni walked out of this London conference when it passed a resolution urging Italian Socialists to dissociate themselves from PCI (*New York Times,* March 22, 1948).

permit Italy to starve, James Carey, then secretary-treasurer of the CIO, announced that though the CIO supported the Marshall Plan, it would not support aid to a Communist Italy. The Italian-American Labor Council, sending $50,000 to Italy, informed the Italians that this was neither government nor Wall Street money, but "it is the money of the rank and file of organized labor."[83] The CIO and AF of L leadership approved a cable drafted by the Italian-American Labor Council warning against extremist "maneuvers and tricks." This cable went to all the daily newspapers in Italy on April 13.

Another aspect of the direct, simple, and open category of propaganda was the letter-writing campaign. There are conflicting claims as to its origin, but it seems probable that Generoso Pope initiated it.[84] Whoever was the instigator, several lent their efforts to build this into a major operation. The basic idea was that Italo-Americans would write to friends and relatives in Italy urging them to vote against the extremists and for the moderate parties with whom the United States could continue to deal. Through his paper, *Il Progresso*, Pope suggested specific questions for writers to tell their Italian correspondents to ask PCI members. V. L. Anfuso, a New York lawyer, went considerably further. He had thousands of form letters printed and obtained the assistance of his church and some fifty Italo-American organizations in distributing them. Monsignor Ottavio Silvestri distributed 10,000 on Easter Sunday at St. Joseph's Church in New York. Priests in "Italian" parishes in Jersey City, Syracuse, Rome (N.Y.), Elmira, and Utica were active in promoting letter-writing among their parishioners. Various social and political Italo-American groups were also active in this campaign. One had over half a million illustrated postcards printed, graphically pointing out the need for democracy to triumph over dictatorship, peace over war, etc. One man, James Hasser, Hollywood producer, paid for his employees' cables to friends in Italy. The underlying idea was that Italians not only would welcome word from American friends, but that they might be more apt to heed the

[83] *New York Times*, April 14, 1948.

[84] See E. Martinez and E. A. Suchman, "Letters from America and the 1948 Elections in Italy," *Public Opinion Quarterly*, XIV (Spring, 1950), 111–25.

advice, or at least recognize the concern, of a friend than they would to respond to less-personal appeals such as the radio, etc. The themes used were along the lines of Christianity versus godlessness; democracy, peace, and happiness versus dictatorship, war, and tears; the end of U.S. aid and slavery to Russia if the PCI won.

Various "stunts" were used to give the program greater impetus here and greater publicity abroad. One was the "Freedom Flights" sponsored by the New England Marshall Plan Committee with the co-operation of the Post Office Department and Trans World Airlines. These flights left on March 26 and April 2, 9, and 16, loaded with airmail to Italy from New Englanders of Italian descent. In a speech in the House of Representatives, Speaker Joseph Martin urged all Americans of Italian extraction to write. He said, "A fifteen-cent stamp might turn the tide for peace." What H. K. Smith says of cables is only slightly less true of airmail letters. "For an Italian peasant a telegram from anywhere is a wondrous thing; and a cable from the terrestrial paradise of America is not lightly to be disregarded."[85] Cortesi reported from Italy that the "Communist-socialist complaint [to the Inter-Party Electoral Committee that U.S. letters were unfair] indicates that the letters from the United States have been effective in opening the eyes of Italians to the reality of the situation. Many understood from them for the first time that the United States means what it says when it declares that American aid will cease if Italy votes communist."[86]

Two impelling themes of what might be called "threat propaganda" originated with Washington. The threat to cut off American aid to an Italy gone communist has been mentioned above. Perhaps even more telling with the individual among the Italian poor was the announcement that the United States would prevent the possible future immigration of any

85 *Op. cit.*, p. 206.

86 *New York Times*, March 23 and April 1, 1948. The effectiveness of this campaign is difficult to measure or to isolate. Probably several million pieces of this mail were dispatched. The New York Post Office reported on March 22 that airmail to Italy had more than doubled in the preceding few days and reached a peak of 1,900 pounds in one day. This sort of campaign might also have boomeranged if Italians had been persuaded that Americans were "interfering," or if the PCI had been able to make this appear as a case where the wealthy uncle was trying to run the lives of his poor relations.

Italian who voted Communist. This was especially thought-provoking among the peasants of the South, in whom the dream of emigration to America is never quite dead. A DOXA survey in April, 1948, showed that over 29 per cent of all Italians would emigrate permanently if they could and another 20 per cent would do so temporarily. Thirty-five per cent of those who would emigrate preferred the United States as a possible home.

Less important, but perhaps cumulatively influential, were such things as individual efforts on the part of tourists[87] and the various displays of American military strength. United States cruisers were almost "stationed" in Trieste. Most of them made individual ten-day visits, but the harbor was rarely devoid of one or two U.S. naval units, since a "replacement" almost always arrived at about the date of sailing of one whose visit was over. Naples, particularly, but Genoa, Spezia, Leghorn, Palermo, Taranto, Bari, and Venice also were frequently treated to the sight of American warships at anchor in their harbors and American sailors on spending sprees in their streets. Sailors ashore are not in all cases good propagandists, but it must be remembered that Italians were used to American servicemen on leave and usually welcomed their friendly, easy-going, spendthrift ways. The importance of the naval units' presence, however, was the implied reminder of American strength in the Mediterranean Sea. In Trieste, on Army Day, April 6, the U.S. contingent put on a parade with tanks and artillery very much in evidence. On the thirteenth, four hundred U.S. Army replacements arrived there by sea, where, because the harbor is the center of attraction, most of the *Triestini* witnessed their debarkation. In the past, replacements had always arrived overland from Leghorn with little fanfare. None of these items may seem impressive, but it must be remembered that they were not actions undertaken in isolation. They were all of one pattern with continuing U.S. aid and the propaganda directly related to that aid.

[87] R. W. van de Velde was in Italy for a ten-day period shortly before the election and knows firsthand the feeling many American tourists had of participating in an "American" election. There seemed to be but little reticence in bars, restaurants, trains, and hotels among Americans quietly electioneering against the FDP.

The Marshall Plan was the single most effective imagination-catcher in the area of aid. Congressional hearings, often more closely followed abroad than at home,[88] commenced on January 8, 1948. On March 14, the Senate passed the plan, later called the European Recovery Program (ERP), and on March 31 the House passed it. The attention paid the plan and its congressional birth pangs may be judged by a Roper poll conducted in the early spring of 1948.[89] The question asked was whether or not the interviewee had read or heard "anything about a plan for American aid to Europe called the Marshall Plan." About 75 per cent[90] of all Italians had read or heard of it. As might be expected, the higher the economic class, the more people had read or heard of it. This would tend to lend weight to Cortesi's statement that the poor did not listen to VOA as they had few radios. Ninety-one per cent of the top

TABLE 4

	Employers	Farmers	Farm Laborers	Craftsmen	Workers	White-Collar Workers	Executives
For ERP.......	80%	76%	47%	64%	47%	76%	95%
Against ERP....	4	9	18	25	32	10	0
No opinion.....	16	15	35	11	21	14	5

quarter (economically) were aware of the Marshall Plan, while only 48 per cent of the lowest quarter had heard of it. Of those who had heard of it, 65 per cent were favorable to it.

Americans might be surprised to find that the Marshall Plan was least popular with Italy's most depressed classes and most popular with its two most economically secure groups (see Table 4).[91] A comparison of the occupational breakdown and one based on political affiliation is revealing of both the politico-economic relationship of classes and the East-West divi-

88 A fact which appears to escape many congressmen when foreign policy is under study.

89 E. Roper, "Where Stands Freedom?" (New York, 1948), booklet published by *Time*.

90 Roper (*ibid.*) reported 74 per cent, but DOXA Bulletin No. 3 (February, 1948) reported 78 per cent. This was less than in France (90 per cent) and Britain (80 per cent) but more than Germany (69 per cent).

91 DOXA Bulletin No. 3, February, 1948. This is the sort of data essential for developing susceptibility and cruciality intelligence.

sion in Italian thinking (see Table 5). DOXA pointed out that political partisanship was the real factor determining attitude toward ERP. Beside this "all others become definitely secondary." More than half of those who favored the plan gave as a reason, "It will help reconstruct Italy," while 71 per cent of those "against" said that "the Marshall Plan is a weapon of capitalism; it will make Europe an American colony."[92] In this connection it should be emphasized that while the United States figured very prominently in Italian minds because of its aid and other activities, the U.S.S.R. was not so prominent. The campaign involved the United States directly, while, to most Italians, the U.S.S.R. took no part. The reason for this was that the FDP, and particularly the Communists who financed and really controlled it, purposely kept Russia out of the picture and even more than that, attempted to keep indications of Cominform-type communism under wraps. There was little

TABLE 5

	PCI	PSI	PSLI	PDC	Right	Independent
For ERP........	6%	33%	83%	78%	86%	65%
Against ERP.....	76	34	4	1	3	4
No opinion......	18	33	13	21	11	31

mention of Russia, which recent prisoners of war and families of men still held there had reason to dislike. Tito was not mentioned either, since he represented the antithesis of Italian hopes for Trieste.[93] Peace, freedom, and ultra-nationalism were the keynotes of the PCI campaign; red flags, the hammer and sickle, and the other usual outward signs of communism were almost totally lacking.[94]

Marshall Plan aid was undeniably an influence in the election campaign. *Il Nuovo Corriere della Sera,* one of Italy's foremost papers *di informazione,* in a subhead on page 1 of March 16, said, " 'The Marshall Plan is one of the most noble

[92] For an Italian Communist view of the Marshall Plan, cf. A. Presenti, "Aspetti politici del Piano Marshall," *Il Ponte,* IV (March, 1948), 210–17. For the opposite view, cf. A. Spinelli, "La ricostruzione europea secondo il piano americano," *Il Ponte,* IV (March, 1948), 193–209.

[93] N. Standen, "Italian Backdrop," *Harper's Magazine,* CXCVI (February, 1948), 150–59.

[94] See Cortesi and C. L. Sulzberger in the *New York Times,* April 11 and 12, 1948, respectively.

and unselfish undertakings which history records,' says the Italian Foreign Minister. . . ." On March 20 a page 1, four-column headline said "With its vote Italy will decide whether or not it accepts the plan for aid."[95] This latter article was a report on Secretary Marshall's speech in California indicating that the United States would not aid a communist Italy since the Communists had rejected the principles of ERP.[96]

While ERP certainly had an impact on Italy, it remained in the realm of ideas during this period. Other aid was, however, a reality and no means of publicizing it was overlooked. On February 2 Ambassador Dunn and Foreign Minister Sforza signed a ten-year treaty of "friendship, commerce and navigation." The fact that this was the first treaty of its kind which the United States had signed since the war was emphasized for Italian consumption. De Gasperi, on February 29 told the nation that aid from the United States had amounted to $1,-869,000,000. PDC posters listed, in various eye-catching graphics, the extent and necessity of American aid.[97] Ambassador Dunn said, at a ceremony greeting the four hundredth ship bearing post-UNRRA supplies, that this was "not a propaganda of vague promises" but a concrete program of action and aid to Italy. He told the assembled audience at Reggio Calabria, an audience largely made up of communist dock workers and their families, that 4,000,000 tons of American grain, coal, and medicines had come to Italy in the preceding six months and he explained to them how the "Lire Fund" worked.[98] On April 3, the active Ambassador was greeting the five hundredth relief supply ship at Taranto. These ships had been arriving at the rate of seventy a week during the preceding seven months and Mr. Dunn told his listeners that the six

[95] Cf. also similar headline articles in *Il Corriere* for March 27, 31 and April 2, 3, 4, 6, and 8, 1948.

[96] Cf. *New York Times,* March 20, 1948. This speech was "counter-counter-propaganda." The FDP had said that America must—could not afford not to —aid Italy, no matter who won the election.

[97] *New York Times,* March 3, 1948, reported that interim aid had been $47,821,505 and post-UNRRA relief was $16,140,883. Cf. U.S. Department of State, *Bulletin,* Vol. XVIII (February 1, 1948).

[98] The sale of U.S. supplies by the Italian government to Italians built up this fund, which was, in turn, used for various relief and reconstruction projects in Italy.

hundredth would arrive in two weeks[99] and that "there are more to come." He totaled up the score to date at 3,250,000 tons of coal, 850,000 tons of grain, 12,000 tons of dried vegetables, 150,000 gallons of DDT—plus fats, evaporated milk, fish, seeds, soap, and medicines including 211,000 units of streptomycin and 90 billion units of penicillin. The total official aid program for Italy for the first three months of 1948 was set at $181,000,000.[100] Mr. Dunn also spoke at Palermo at ceremonies marking the arrival of a U.S. supply ship and at Genoa at the celebration, on March 23, of the start of the project to refit two Italian ocean liners. Here he again pointed out to the workers the direct way they benefited from U.S. aid. The refitting of these two ships was possible because of the "Lire Fund" from sale of American supplies. On April 2 Ambassador Dunn handed the Italian Minister of the Treasury a check for $4,300,000 as partial payment for sums due Italian prisoners of war who had worked voluntarily for the United States while in captivity. The debt, to 60,000 men, was to have been paid by Italy, but America undertook to relieve Italy of that burden.[101]

All aid was not official. American Relief for Italy, Inc., sent over 2,000 tons of supplies, including 323 cattle, in January and February. And the War Relief Services of the National Catholic Welfare Conference had sent 27,000 tons of relief goods since October, 1944.[102] American labor's gift has been mentioned. Myron C. Taylor made a gift of $75,000 to Premier De Gasperi for work in caring for abandoned children and children of wounded veterans.[103] Apparently gift parcels from individuals to individuals also played a part in the campaign. Don Sturzo, writing in *Il Popolo* said, "Those who receive gift packages from uncles and cousins in the United

[99] He met it in Naples on April 15 and Dunn's speech was the first in which he clearly alluded to the elections. Previously he had confined his remarks to what aid was arriving and how it was helping Italy (*New York Times*, April 16, 1948).

[100] U.S. Department of State, *Bulletin*, XVIII (February 22, 1948), 236. Most of it was allocated to food ($106,000,000) and fuel ($59,000,000).

[101] *New York Times*, March 24 and April 3, 1948. On March 28, Cortesi reported that Ambssador Dunn's speeches were "widely reproduced by all but the Extreme Left-wing newspapers" and that he had done "much good work in driving home to Italians what American aid means to them."

[102] *New York Times*, March 11 and April 14, 1948.

[103] *Il Corriere*, April 8, gave page 1 prominence to this gift.

States have been warned that they will not get another dollar if Italy turns Communist. . . ."[104] Another "stunt" which combined unofficial aid with propaganda was the "Friendship Train" which had toured the United States gathering gifts and then toured Italy distributing them. The cars were brightly painted red, white, and blue, and large signs on their sides explained their mission, emphasizing the friendship of American citizens toward the citizens of Italy.

All the propaganda was not carried out in Italy. President Truman, at a White House ceremony with Italian Ambassador Tarchiani as guest, gave Italy twenty-nine merchant ships on March 16. Fourteen of these were Italian vessels seized here during the war and fifteen were Liberty Ships equal in tonnage to Italian ships seized by the Allies and later lost while in Allied service. But it should not be thought that Italians were entirely concerned with the direct aid they could get from the Americans. They heard about the ships, but the biggest headline in *Il Corriere* the next morning was about President Truman's "emergency measures."[105] The following day, March 18, *Il Corriere* carried a six-column, page 1 headline saying, "The United States will stand beside the nations of Europe in the defense of their liberty." There was also a three-column picture of Mr. Truman addressing the joint session of Congress.

OTHER ACTIONS HAVING A SYMBOLIC EFFECT

This brings us directly to the third type of American influence in the Italian elections—actions or policies which, while per-

104 Quoted in the *New York Times,* April 1, 1948. Cf. DOXA Bulletin Nos. 5, 6, April, 1949, which, while almost a year after the election, probably shed some light on the situation relative to individual aid in 1948 and the familial and sentimental ties of Italians with nationals of other countries. Sixteen per cent of the Italians claimed relatives in the United States. (DOXA believes these are only those still "in touch"; that actually there are many more.) Almost 24 per cent of all Italians have received gifts from abroad "in the last few years"; 64 per cent of these from the United States, none from the U.S.S.R. Gifts came from relatives (82.6 per cent), friends (8.2 per cent), UNRRA (4.4 per cent), CARE, etc. (0.6 per cent), other charities (4.2 per cent). It must be remembered that these were direct, personal gifts to the individual concerned and not "aid" indirectly beneficial.

105 *Il Corriere,* March 17, 1948. The measures were the President's proposals to link inflation curbs, UMT, the draft, and foreign aid to meet the crisis caused by the continued political aggression of the U.S.S.R. in Europe. The Russians were also then beginning to try to pinch the West out of Berlin.

haps directed toward another nation, were either designed or timed to have a psychological effect in Italy. As has been said, the United States was very evidently in the campaign, whereas Soviet Russia had remained at least superficially aloof. There was some feeling that this appeared to make the United States a busybody or an "imperialist power," depending on one's point of view, while the Russians remained correctly detached by operating through the Front or the PCI, which, in turn, was doing everything possible to minimize its connection with the Soviet and the Cominform. The problem then was to make the PCI (and therefore the FDP) admit its international connections and to make Russia reveal to the Italians its real interest in the matter. Two issues were at hand and very adroitly used by the West, largely at American instigation. Later facts proved that the Kremlin-Tito break was well along at this time, but the news was not known abroad and it was believed the Russians would not agree to Trieste's return to Italy. This was a shrewd gamble and it worked. It proved to be of great propaganda value. On March 20, through M. Bidault, who was meeting with Italians on the matter of a French-Italian economic accord, the three Great Powers of the West announced that they had proposed to Russia that she agree to the return of the Free Territory of Trieste to Italy, as it was apparent no other arrangement was going to be reached and that the sentiment in Trieste was mostly Italian.[106]

The reaction in Italy was immediate. Even a Communist-ordered printers' strike could not prevent the spread of this good news, although it probably did slow it down. *Il Corriere*, like all dailies which had the news before the strike could be called, printed it. In a six-column, page 1 headline it said that this Anglo-French-American proposal was the "first step in the revision of the treaty." And in Milan, *Il Corriere*'s hometown, but also usually a stronghold of communism, on March 21 the populace was so hostile to a large group of communists staging an anti–Western democracies rally, that the Reds were obliged to seek safety in flight. After various Russian mutterings about the sanctity of treaties, and Tito's offer to swap Gorizia for Trieste, the Kremlin on April 13 turned down the

[106] *New York Times*, March 21, 1948. Cortesi reported the news was "received with unbounded joy in Rome" and might "considerably influence voters against the Communists."

West's proposal. Meanwhile *Il Corriere* had several times made front-page news of the issue and De Gasperi had used it to challenge Togliatti to support Italian claims. The result was embarrassing to PCI, since they were forced into openly defending the Russian-Yugoslav side. On April 15, *Il Corriere* in a front-page, five-column headline said, "Moscow's 'No' on Trieste falls at the feet of the Italian Communists." On the eighth the Allied Military Government in Trieste announced that it would turn over to the civilian authorities on April 12 all the city's administrative functions, except for that of public safety. Such a move was normal and probably fully justified by existing circumstances, but the timing and publicity, contrasting so sharply the difference between Western and Russian attitudes, can hardly be attributed to coincidence.

Another issue close to Italian hearts which was used psychologically against the Front and Russian prestige in Italy was that of Italy's membership in the United Nations. On April 3, the United States, the United Kingdom, and France proposed once more that Italy be admitted to the United Nations. This was again obviously a means of forcing Russia to disclose her attitude toward Italy (she had twice previously vetoed the latter's admission) and a method of embarrassing the FDP by making it unable to support Italian aspirations and still remain pro-Cominform. The April 4 issue of *Il Corriere* carried the story on page 1 under a four-column heading, "Admission of Italy to the UN asked by the United States, Great Britain, and France." Again on the eighth the paper made a front-page story of Russia's apparent decision to veto the proposal. On the tenth the Western Allies forced the matter to a vote, and Russia, acting as expected, vetoed Italy's entry for the third time. This time *Il Corriere* used a five-column front-page headline to announce the U.S.S.R.'s latest show of unfriendliness. It is difficult to find reasons for Russia's apparent abandonment of her Italian satellites and her obvious disservice to them in their effort to turn the country Communist. Whatever the reasons may have been, the results were (1) to reveal the falsity of PCI claims to being supporters of Italy vis-à-vis the world at large and (2) to solidify and strengthen an otherwise amorphous "Center" in Italian politics.[107]

[107] Cf. H. K. Smith, pp. 203–4; and L. Valiani, "Sulla genesi della 'terza forza,'" *Il Ponte*, IV (February, 1948), 104–8.

"ACCIDENTALLY" SYMBOLIC ACTIVITIES

There are two or three manifestations of American policies and actions which fall in the fourth kind of influence mentioned earlier. These were "accidentally" symbolic insofar as Italians were concerned. Chronologically the first was the U.S. House of Representatives' addition on March 30 of Spain to those nations eligible to participate in ERP. This, whatever else it may have been,[108] was very poor propaganda for the United States in Italy as well as in Western Europe generally. James Reston, writing in the *New York Times* of March 31, said that this move gave basis to the Italian Communist propaganda that the United States was reactionary. If any proof were needed, it was furnished the next day when Saragat and Lombardo cabled various U.S. labor organizations indicating that PSLI was "profoundly perturbed" by the inclusion of "Fascist Spain" and saw in this action a "great threat to the cause of European democracy."[109] Further substantiation of Reston's claim was to be found in the article, obviously happily printed by *Unità*, the PCI paper, stating that here was proof that the Marshall Plan "was not made to support democracy nor to restore freedom."[110] *Il Corriere*, while reporting the House action, apparently played it down by captioning it only as a subheadline on page 1 under a large headline on Myron Taylor. Perhaps with a sense of relief that paper made a two-column, page 1 headline on April 2 of the fact that President Truman had excluded Spain from ERP. Actually, to emphasize the point, the administration directed Ambassador Dunn to make a formal call at the Italian Foreign Office to announce Spain's non-participation.

The U.S. stand in Berlin against Russian blockade tactics perhaps should not be classed as "accidental symbolism." Certainly one reason for our firmness was to prove to the rest of Europe that we did not intend to be forced out by the U.S.S.R.

108 *New York Times*, March 31, 1948, reported that U.S. military heads were pleased but that the State Department was embarrassed.

109 *New York Times*, April 1, 1948. Ivan Lombardo, a long-time and well-known Socialist leader, had quit PSI and joined PSLI in February, 1948. With him had gone Ignazio Silone, prominent Socialist author. These "conversions" away from the extreme Left may have been partially influenced by America.

110 *Unità*, March 31, 1948.

and, consequently, to strengthen the determination of free Europe to resist "internal aggression."[111] Nevertheless, the Russian blockade was a surprise and therefore the results from our countermoves cannot have been planned in advance. It is interesting that throughout April *the* news story which received most attention on *Il Corriere*'s front page was the Berlin blockade, and it is particularly noteworthy that the "slant," if it can be said to have had one, was the play-up of America's firm resistance. The general tenor indicates that America's attitude not only surprised the editors but greatly pleased them. A sidelight of the blockade-counteraction story which also appears to have impressed *Il Corriere*'s editors was the fact that U.S. naval units were sent in early April to visit Norwegian ports. This was interpreted as a gentle warning to the U.S.S.R. not to harbor any ambitions in the direction of Scandinavia. There are interesting parallels to the case of Italy and the Mediterranean which apparently were not lost, at least on Italian editors. Norway and Italy both have very long coast lines in waters formerly dominated by the British Navy, but now apparently largely under U.S. Navy surveillance. The American Navy had been used for some time as a reminder to Italians to avoid capitulation to an internal communist force; it was now being used to remind the fountainhead of communism not to encroach further toward the West.

The campaign officially closed at midnight April 16, allowing one day for cooling-off before the actual polling on April 18 and 19. Each side had claimed that the other was a tool of a foreign power, and all the issues which would normally have called for various shades of gray in public opinion were simplified and magnified in stark black or white. By the end of the campaign there were no "maybes" or "perhapses" or "yes, butses"; only a direct "yes" or "no" appeared adequate. One Italian political writer summed it up thus:

> The central argument of every discussion; *Communism* or *anti-communism*. . . . In the constitutional field a choice between liberty or dictatorship; in the spiritual field, between salvation and damnation; in the economic field between bread and hunger; in the international field, between America and Russia, and even between

111 See L. D. Clay, *Decision in Germany* (New York, 1950), *passim.*

war and peace. . . . And it is curious that everywhere and on either side of the quarrel one heard the same appeal: "Liberty and peace are on my side; he who votes against me is for dictatorship and war. . . ." And the other side used the same phrases like an echo.[112]

Opposing Propaganda Appeals

Three weeks before the election the *New York Times* Rome correspondent summed up[113] the main points of the Communist and anti-Communist propaganda. The Communists claimed the following: (1) De Gasperi has sold out to the United States; this will inevitably end in war for which America is preparing. The U.S.S.R. stands for peace. (2) Only a "people's government" [extreme Left] can carry out the agricultural and economic reforms needed. (3) American aid to Italy will continue no matter who wins, because America has overproduced and needs Italy as an outlet. The presidential candidates [1948 U.S. elections] will need the Italo-American vote in November. (4) The Vatican, chief backer of PDC, is a center for graft and the black market. The FDP will guarantee religious freedom but will tighten up controls on such Vatican activities. (5) The West can't give back or get the U.S.S.R. to give back Trieste. Only PCI, through direct negotiation with Tito, can get it back. (6) PDC and De Gasperi are hurting Italy and the working class by sabotaging relations with Tito and trade with Russia. Russia can give all that the United States can, and under better conditions.

It should be noted how difficult this propaganda would be for PDC or any *Italians* to refute. Some could only be countered by the Vatican; most of it only by the United States or the U.S.S.R. It is interesting how thoroughly American propaganda undertook to "pull the rug out from under" these propaganda lines. It emphatically and repeatedly, through official and unofficial sources, stated that American aid would cease if Italy voted the Front into power. It forced into clear highlights the attitude of Russia on Trieste and her attitude toward acceping Italy as an equal member in the family of nations. Indirectly it caused Tito to expose his attitude toward Trieste in

112 P. Calamandrei, "Preludio al 18 Aprile," *Il Ponte*, IV (May, 1948), 406.

113 Arnaldo Cortesi, *New York Times*, March 28, 1948.

the unacceptable deal he offered to make in regard to that city. And of course by continuing, even speeding up aid and publicizing it, America underlined the fact that no aid had come from Russia, only demands.

Conversely, the reader will readily see how U.S. propaganda and actions strengthened and gave substance to several of the anti-Communist claims which Cortesi summarized[114] as follows: (1) Communism entails a police state. Italians fought for freedom and must not now give it up for "visionary chimeras." (2) A Communist victory would spell ruin for Italy. Free American aid would cease. Italian industry, which obtains 90 per cent of its coal from America, would come to a stop. (3) Starvation would follow a Communist victory. The present ration level is maintained only through American aid. Conditions behind the Iron Curtain are bad; the Communists could not help Italy even if they would. (4) A Communist victory would end cultural and commercial relations with the West. (5) Russian expansion is the main threat to peace. If Italy falls behind the Iron Curtain, her chances for peace would be slim. (6) The existing government with the West's aid has helped Italy regain equality in the world. The treaty may be revised. Russia has vetoed Italy's admission to the UN, opposes a revision of the treaty, and insists on $100,000,000 reparations from Italy.

As usual in a propaganda battle, each side had some truth in its claims. The government had been very slow in instituting reforms, particularly in agriculture and land tenure. Workers in industry were also in a bad economic plight. The United States was obviously interfering in Italian politics, as was the Vatican. These were facts the PDC could not hide; it could only hope to minimize them and to emphasize the improbability of the Front's capability for improving the Italian situation. In this, all evidence indicates, U.S. influence was certainly a major factor even if there is no method of "measuring" it.

114 *New York Times*, April 3, 1948.

• CHAPTER VII

Radio Free Europe's Operations VETO and FOCUS

Radio Free Europe was founded in 1949 by a group of private American citizens.[1] It has twenty-nine powerful transmitters located in West Germany and Portugal with which it broadcasts with completely independent program schedules to five satellite countries behind the Iron Curtain—Poland, Czechoslovakia, Hungary, Rumania, and Bulgaria. Unlike the official Western stations like the Voice of America and the British Broadcasting Corporation, RFE is the independent voice of exiles which tries to present a genuinely patriotic "home service" radio for the peoples in the five captive nations behind the Curtain.

From the beginning, of course, it has had political goals—the most significant being the ambitious one of trying to contribute to the peaceful liberation of the satellites. In the first three years of its existence there was very little realistic concern with how RFE could contribute to this goal. Reading some of the policy guidances, one gets the impression that the only strategy involved some hazy notion of "drilling" holes in the Iron Curtain through which "Truth" could be "poured"—and it was known from biblical authority that "the truth shall make them free." But by the beginning of 1953 there was some

[1] This chapter is adapted from Robert T. Holt, *Radio Free Europe* (copyright, University of Minnesota, 1958) and is printed here with the permission of copyright holder.

hard thinking about strategy, followed by a certain amount of frustration. RFE had forged a mighty technical weapon with which it could pierce the Iron Curtain. Its goal of liberation had the official support of the new administration in Washington. But how could it use its instrument to achieve its goal? It could not build up the hope that Western armies would soon come smashing through Eastern Europe. Not only was this contrary to American policy, but in the age of thermonuclear weapons, major war could mean total destruction. No promise of liberation could be appealing if it followed in the wake of atomic annihilation. Nor could RFE incite violence or call for mass uprisings and revolution. The century of the barricade revolution passed with the coming of the machine gun. Even if the appeal were effective, it would bring only a blood bath —not liberation. RFE seemed fated merely to support the morale of the captive peoples and to harass the regimes in its daily broadcasts. What was needed was a strategy—a plan of action to reach the stated objective. But there were two prerequisites to the development of a strategy that had not by this time been met. First, psychological operations can only be successful in achieving a difficult and long-range goal if they are employed as part of a total strategy involving the use of all the instruments of statecraft. The U.S. government had a goal of liberation, but no strategy of liberation. If one had existed, RFE could have used its facilities to help implement it. In its absence, RFE either had to be content to broadcast with little purpose or develop at least the major outlines of a strategy of peaceful liberation. Second, there had to be a thorough analysis of the situation to determine what kind of behavior would be feasible for the people behind the Curtain to undertake. RFE's information divisions and policy advisers began to lay the basis for meeting the second prerequisite following the Czech and East German riots of June, 1953.

The Events of June, 1953

In Czechoslovakia, the youngest of the People's Democracies, 1952 ended on a note of terror. Slansky, Clementis, and nine other Communists were tried on a number of charges ranging from ideological deviation to sabotaging the economic plan. They were, of course, found guilty and executed, and the whole affair was used as a pretext for introducing new and harsher economic policies. In the spring the repressive measures were intensified by Zapotocky (who took over the reigns of government when Gottwald died shortly after returning from Stalin's funeral). The most drastic step was taken on May 30, when rationing was abolished and a currency reform instituted. This action swept away savings, depressed the already low standards of living, and forced workers to increase production by keeping them on the job longer.

But at this point something happened which the police state fears but rarely has to deal with. On June 1, large-scale riots broke out at the famous Skoda works in Pilsen and rapidly spread to a number of other cities. The workers were protesting the new reform. The demonstrations were violent but short-lived; on June 15, the regime was able to sponsor a large workers' rally in Bratislava which passed off without incident. The stringent economic course was (for the time being) maintained. But something happened that neither the regime nor the people could forget, although its full import was not evident until sometime later.

While the Communist oppression was being increased in Czechoslovakia, the old policies were suddenly reversed in East Germany. The collectivization drive was curtailed, workers' norms were lowered, the investment rate in heavy industry reduced, and an "agreement" was signed with the Evangelical Church. However, the people were not in a grateful mood. They had caught the glimpse of a better life and rose—in what was almost a rebellion—to force the regime into even greater concessions. The uprising started in Berlin and rapidly spread to many other places in the East Zone. They were only brought under control when the Red Army moved in—in full battle array. Stones would not stop tanks, and rioters were brought under control.

The Czech and the German demonstrations developed out of entirely different situations. The Czech workers rioted when they felt themselves being pushed deeper into the trough of misery; the Germans interpreted the granting of concessions as signs of weakness and they rose to push their advantage. But there was one all-important similarity. In neither case were the demonstrators put down with the ruthless brutality which

RFE had expected would follow acts of overt opposition. Tanks moved into Berlin streets, but, by and large, their guns remained silent. Arrests were surprisingly few, mass executions unknown.

RFE's interpretation of these events had three major points: First, and most important, there exists behind the Iron Curtain a large opposition. It is unorganized; it is inarticulate. But in Czechoslovakia (where our interest is focused for the moment) it included large sections of the working class and of the youth who in 1948 supplied the mass support for the Communist party. It even included members of the Communist party. Some shop stewards at Pilsen threw in their lot with their fellow workers. Second, the rulers were disorganized and uncertain of themselves. Their leader of thirty years was dead. They were not sure of the new policy of the Kremlin; they did not know what course the power struggle might take. They hesitated to move vigorously against the rioters, not because they had renounced brutality, but because they had lost confidence—not because they were more merciful, but because they needed the workers in order to meet the production demands made by the Kremlin. The third point follows from the first two. The workers, if they stood firm, could make demands on the regime that the regime would have to honor. These three propositions arising out of an analysis of the June demonstrations underlay the VETO and FOCUS operations which were started the following year. One can see in this interpretation how the criteria of susceptibility and cruciality were taken into consideration. Large sections of the working class and some youth groups were susceptible because they were motivated to act in certain ways and because the regime was not prepared to use brutality to prevent some kinds of activities. The workers were crucial because the leaders needed them if they were to meet the production demands made by the Kremlin. This analysis in turn contributed to the assumption about the feasibility of the goals of liberation. If all the ramifications of a cruciality, susceptibility, and feasibility analysis spelled out in chapters ii and iii had been taken into consideration, one wonders if a different interpretation might have been made.

After a few days of initial confusion RFE developed a new propaganda line for its broadcasts to Czechoslovakia. Its purpose was to convince the people—most of all the industrial workers—to use their own strength and ingenuity to begin hacking away at the Iron Curtain from the inside.[2] Radio Free Europe's policy guidance for workers' programs called for scripts which would make the workers aware of the significance of the events in Pilsen and Berlin. It was crucial to convince the industrial workers that their strength was increasing while the rulers were confused and disorganized.[3]

In speaking to the police and to the army, RFE tried to undermine any loyalty that they might have to the regime—to make them unreliable instruments of control, particularly in times of crisis. Broadcasts to the police were to follow the line suggested by this paragraph:

> Is it true under a true people's regime or under a reactionary regime that the police fire upon the oppressed and striking workers? What is the mark of a Fascist? Is it not that he takes sadistic pleasure from shooting and beating the workers in his own country? Must a man be hated by his neighbors simply because he wears the uniform of a policeman?[4]

To the army the questions were similar. Radio Free Europe wanted to remind the officers that they owed allegiance not to the Communist leaders but to their country and to their people. The rank and file were told that they might some day be asked to shoot at their brothers in the streets of their own cities. Radio Free Europe had reported that when the German *Volkspolizei* were called into the street to suppress rioters, they were something more than embarrassed. They held their fire because they could not shoot down their brothers. Some even put their guns into the hands of the workers. And the Czech and Slovak soldier was asked, "If that day should come for you . . . would you do less than the East German *Vopos* did?"[5]

Next to the workers, officials in the party and in the regime were the most important targets in RFE's revised campaign. Officials were reminded that chances of their living to a ripe

[2] "Czechoslovakia—Guidance No. 10," June 30, 1953, p. 2.

[3] *Ibid.*, p. 8.

[4] *Ibid.*, p. 5.

[5] *Ibid.*, pp. 5–6.

old age were not great. Most servants of the Kremlin, great or small, live to be disgraced, purged, imprisoned, perhaps shot. The risks were even greater during the period of uncertainty when they did not know what turn the power struggle in the Kremlin might take. In a situation like this, party loyalty means nothing and personal loyalty is dangerous because one never knows when the object of that loyalty in the Kremlin may fall. And all the time the officials are known by the people to be agents of a foreign power. When the people are free, the officials may expect to be torn to pieces, as was the case with the collaborators with the Nazis in France in 1945.[6]

The appeal to the officials was concentrated on giving them two action alternatives. Either they could cross the "golden bridge" of forgiveness by the people and use their positions to sabotage the Muscovites and protect the people in that part of the nation under their charge, or they could defect to the West.[7]

And finally there were the youth groups. For three years RFE had been addressing the students and telling them not to manifest hostility against the regime—not to show their disbelief at the preposterous Communist teachings—not to cry out that a "poem written by an illiterate sycophant to celebrate the manufacture of a tractor is not literature." Rather the student was told to be quiet and learn what he could from his parents. He was told to learn a profession, to learn to handle firearms, to handle a platoon of soldiers, because someday his country might need his skills. Following the June riots they were told in addition:

> Be suspicious. Do not be taken in by strangers who talk about resistance. Be not tempted by the promise of adventure, of secret meetings. Do not look for leaders. But at the same time, think about what you can do, what part will be yours when the time comes. And when you think, remember how great will be the reward—to be a free man rebuilding a free country in a free, fraternal United Europe; to have a Koruno that will actually buy something—an English motor-bike, a French literary review, a novel by Hemingway, a honeymoon on the Italian lakes, all in a world without police. It is

6 *Ibid.*, pp. 9–10.

7 *Ibid.*, p. 10. This is one of the few instances in which RFE supports defection.

you who will rebuild our society according to your ideas, not a foreigner's ideas, of what the life of our country should be.[8]

These themes suggested action but they did not yet constitute a strategy because they were not related to any obtainable goal. The plan of action in Czechoslovakia (VETO) did not start until the following year. In order to understand how it was developed, it is necessary to review events behind the Iron Curtain in the summer and fall of 1953 and RFE's response to them.

The New Course

In the first days of July, 1953, Nagy replaced Rakosi as head of the Hungarian government. Four days later he repudiated some of the stringent policies of the old regime and launched Hungary on the "New Course." In August it became evident that the "investment in moderation" was spreading to Czechoslovakia when President Zapotocky criticized the policy of rapid forced collectivization and announced that peasants would be free to leave the collectives. Although the New Course went further in Hungary than in the other satellites, we will trace its development in Czechoslovakia because of the close relationship between the unfolding of events there and the development of VETO.

The New Course really got under way in Czechoslovakia in the middle of September when the regime announced a new economic program which included the following points:

a) Curtailment of the investment rate in heavy industry
b) Greater investment in all sectors of agriculture and higher prices allowed peasants for grain
c) Relaxing of the pressures that had been exerted against the independent (non-collectivized) peasant
d) Increase in quality and quantity of consumer goods
e) Augmentation of the housing program
f) Broadening of the raw-material base, specifically in order to increase production of thermal and electric power

And it was soon evident that the program was not just an empty propaganda gesture. By the end of September prices were reduced on 23,000 consumer items. In November new

[8] *Ibid.*, pp. 6–7.

assistance programs for both the collective and the independent peasant were announced. In December food prices were cut, income taxes favorably revised, buying restrictions on some consumer items reduced, and delivery quotas on many agricultural products lowered.

The New Course involved not only economic measures; speeches made in December by Siroky and Novotny made it clear that it also contained important political and ideological elements. The following schedule of political events was announced to take place in the late winter and early spring:

a) Elections to party committees of lower rank in February.
b) Elections to shop committees in late February or early March.
c) Elections to local, district, city, and regional national committees, probably in May. (These committees came spontaneously into existence during the liberation of Czechoslovakia in 1945. They were given legal standing and wide powers as organs of local government in the draft constitutions of 1946 and were placed under the Ministry of Interior. Siroky declared in December, 1953, that jurisdiction over them had been transferred to the cabinet.)
d) Parliamentary elections (required by the constitution).
e) Tenth Congress of the Czechoslovak Communist Party, June 11–13.

At the same time that this political program was announced the ideological line was laid down. Siroky and Novotny in their December speeches made it clear that the new policy represented no retreat from basic Marxist principles. In fact, the success of the New Course was predicated on the reaffirmation of these principles. It represented the tactical changes of the party to meet a new situation. It was not a compromise with bourgeois nationalism, social democracy, or "Masarykism," which were singled out for specific criticism—the first because it would open the old sores of Czech-Slovak dissension; the second because it would introduce anti-Marxist concepts of "equalitarianism"; the last because it was merely the old bourgeois capitalism masked behind a fraudulent humanitarianism.

The whole party apparatus would also have to be revitalized. A new spirit of militancy would be needed to uproot functionaries who were living in isolation from the masses and getting by merely by repeating the old slogans. All of the instruments of propaganda and agitation would be used to instil the new enthusiasm among party members and through them to the masses.

This was the situation in January, 1954. The policy advisers at RFE felt there were two main aspects of the New Course. The first was purely economic. The Communist world was beginning to realize that if it were to achieve its goal of catching up economically with the West, it would have to increase productivity as well as production. This could not be done without eliminating the grosser errors and inefficiencies in the whole planning structure. But more important, in the last analysis it meant gaining the cooperation of the workers. Terrorism could enforce high norms and reduce absenteeism and thus increase production; it could not raise productivity (output per man-hour). This could only be done by granting concessions, including more consumer goods. It was hoped that the resultant increased productivity by more satisfied workers would result in the production of both more guns and more butter.[9]

But the concessions could not be granted unless they were strictly controlled. The Communists were determined not to let concessions in Czechoslovakia have the result that they had had in Germany the previous June. And so the second part of the New Course involved a strengthening of the party apparatus, actually to increase the degree to which the masses in their new economic freedom were politically controlled by the regime. New elections were being held to shop committees, because during the June riots the old committeemen had either thrown their lot with the workers and therefore could not be trusted by the regime or they had so earned the contempt of the workers that they no longer had any influence on them. New local party chairmen and secretaries were needed because the old ones were either living in isolation from the masses or had grown so fond of their friends and neighbors that they were protecting them from the decrees of the regime rather than enforcing them. New national committees were needed, because the regime needed a new enthusiastic and trustworthy

9 "Czechoslovak Guidance No. 13: The 1954 Czechoslovak Electoral Campaigns," January 25, 1954, p. 4.

instrument with which it could enforce its will in the countryside.[10]

At the end of January, 1954, RFE began to develop a plan to deal with this new situation.[11] It had three phases. The first was to run during the pre-election period or until about February 20. During this phase RFE would explain to the people of Czechoslovakia the real meaning of the New Course—how the age-old combination of the carrot and the stick was being used to accelerate the process of sovietization. The reasons for the economic concessions and the new election and the relationship between the two would be broadcast across the countryside.

The second phase would run from mid-February until mid-May. During this period RFE would actively engage in the campaigns for election of new shop committees and new national committees as the "Voice of the Opposition." It could not, of course, support candidates of the opposition, but it would present the demands of the people and let the candidates know what the workers expected of them.

The third phase would correspond to the period of the campaign for elections to the parliament. Although RFE felt that the shop committees and national committees could be used by the people for their ends, the parliament could only be a rubber stamp for the regime. Thus, during the third phase its efforts would be devoted to discrediting the parliament and discounting the elections as a shameless travesty on democracy.

Operation VETO

All of the basic tactical elements of the first phase of the VETO operation were contained in this January Plan. VETO, which began in late April, represents a refinement of these elements and a general intensification of activities in an integrated, long-range *action program.* The January Plan laid out a tactical maneuver designed to frustrate the regime's intentions of using the round of elections to revitalize and consolidate the state and party machinery. VETO was a campaign aimed at achieving eventual liberation. The election fight would merely be the first step in a program to return political initiative to

10 *Ibid.*, pp. 4, 7.

11 *Ibid.*, pp. 1–7.

the people. Behind this change from a tactical maneuver to a long-term campaign lay a new estimate of the situation, which included the concept of liberation as a result of a confluence of forces operating within and outside the satellites.[12] The outside force was the strength of the free world; the inside, the strength of the people's opposition. RFE saw its long-range task under Operation VETO as building up the strength and action potential of the internal force against that day when a favorable relationship between the internal and external components combined with an international situation, in which the Soviet Union would find it more painful to intervene than not to intervene, would lead to the liberation of Czechoslovakia.

We stated above that in its first phases VETO represented a refinement of the tactical elements involved in the January Plan and a general intensification of activities. The important refinement was in the concept of the "People's Opposition"; the intensification consisted of the combining and integrating of the radio with the balloon-leaflet campaign.

The realization that there was a large unorganized opposition in Czechoslovakia which was capable of engaging in politically significant action came out of the June riots. The January Plan spoke of RFE as the Voice of the Opposition. The idea was extended to its logical conclusion in Operation VETO. RFE spoke to the people of Czechoslovakia of the existence of a People's Opposition. It was a movement without formal organization, without a headquarters, without material means, but one to which every Czech and Slovak belonged, if he was opposed to the process of sovietization that was taking place in his country. It was not a conspiracy or an underground. Essentially, it was a spiritual union of all patriotic Czechs and Slovaks. In VETO, RFE spoke as the voice of the People's Opposition. The people themselves had to be careful even about what they said in private. No internal voice of the opposition would have the means or the opportunity of commanding a mass audience. RFE tried to fill this gap. It could speak clearly, forcefully, and, it hoped, authoritatively.[13]

12 For a detailed discussion of "confluence of forces" see Holt, *Radio Free Europe*, pp. 20–22.

13 One of the reasons why RFE felt it could assume the role of the "Voice of the Opposition" was that it was much better informed about events in Czechoslovakia than all but a handful of people inside Czechoslovakia.

An effective opposition has a platform and RFE articulated and disseminated the "Ten Demands of the People's Opposition" as its platform. In brief these demands were as follows:

1. Trade Unions for Trade Unionists
2. Better Pay—Less Talk
3. Workers Must Not Be Chained
4. No Meddling with Workers' Free Time
5. No More Farmers' Bondage
6. Smaller Quotas—Larger Crops
7. Local Autonomy instead of Bureaucracy
8. Goods for the People, Not for the Soviets
9. Back to Serving the Customer
10. Housing for Families, Not for the State

These demands were not the dream of some American or exile sitting in New York or Munich reflecting on the situation in Czechoslovakia. These were the demands of a captive people as RFE determined them from numerous information items and a careful reading of the regime press. They existed in Czechoslovakia before they existed in Munich. RFE merely articulated and disseminated them. These demands became the solid substance of Operation VETO.

The development of these Ten Demands is an example of the work of output intelligence. The analysis of newspapers and journals from Czechoslovakia, monitored radio reports, and interviews with refugees indicated that many people in Czechoslovakia would like to express these demands, and it was argued, therefore, that they could be used to make up an effective propaganda message.

VETO marked a refinement in the concept of the Voice of the Opposition: it involved an intensification because of the integrated use of radio and balloon-leaflets. The relative advantages of the spoken and written word have long been recognized by propagandists. The tactical flexibility of radio, its all-pervasive nature, and its ability to capitalize on the emotional power of the human voice are all limited by the transitory nature of any given program and the difficulty of dealing with complex and difficult thoughts. But its weaknesses are precisely the strengths of the written word. The VETO plan was devised to take advantage of the complementary nature of radio and leaflets. Basic messages were carried in by balloons.

They were commented on and interpreted in light of day-to-day events on various programs on the "Voice of Free Czechoslovakia."

In discussing the development of VETO it is useful to look at a number of phases identified by certain specific objectives and themes. Phase 1 was short and introductory. It was designed to capture the interest and stir up the curiosity of the Czechoslovak public and to plant the idea of the Ten Demands as the platform of the People's Opposition. It began late in April, 1954, with the dropping of a sticker containing only the number "10" and with a campaign of radio spot announcements designed to stimulate interest in the number "10." Starting just after the first of May, 1954 (scheduled to begin on May 1 but postponed because of unfavorable winds), the small newssheet *Svobodna Europa* ("Free Europe") listing the Ten Demands and commenting on them was carried into Czechoslovakia by balloons. This publication was dropped until shortly after the middle of May. In the meantime four other leaflets were dropped with the aim of popularizing the Ten Demands and making special appeals to specific groups (farmers, workers, party functionaries, etc.). These leaflets were in the form of stickers and gave the people the means with which they could participate in the electoral campaign by pasting them up in public places. The themes developed in leaflets were also carried in radio programs.[14]

The second phase got underway in the latter part of May. It had two important objectives. The first was to explain exactly what the Ten Demands really meant to individuals; the second and more important task was to try to get the people to understand why the opposition campaign was based on specific and attainable demands (i.e., why an *action program*). The first involved relating the demands to the specific needs of individuals in Czechoslovakia and penetrating so deeply that it would be impossible for them to hear or see "10" without political connotations coming to mind. The balloon-leaflet implementation of this was accomplished by two leaflets. The first was the "Ballot of the People's Opposition" issued in twenty million copies. It carried a statement and exposition of the Ten Demands.

[14] Operation VETO (Phase 1), May, 1954, pp. 2–3.

The second and far more delicate mission of phase 2 was to explain the reason for an *action program* based on the Ten Demands. Why, instead of sustaining the assumption that liberation would come, albeit eventually, from the free world, did RFE begin to speak to the people of Czechoslovakia as essential partners in the struggle? Why, instead of polemics and exhortatory slogans, did RFE now talk about the People's Opposition and its ten specific demands? The answer to these questions involved an exposition of the thinking that lay behind the whole VETO operation. In brief, the reasoning was presented to the Czechs and Slovaks as follows: Liberation as the consequence of an aggressive war waged by the West was unrealistic. Liberation as the result of a reformed Kremlin voluntarily withdrawing from the satellites or liberation as the result of successful Titoism was a fantasy. Viewed seriously,

> . . . liberation of the Soviet satellites is likely to occur only as the result of a *favorable confluence of events within and outside* of the country. Inside this means that *coordinated mass opposition* must evolve and gather strength during the preliberation period. Outside the country it means that the "People's Opposition Center in the Free World" must provide initiative, tangible symbols and (in part) the ideological basis for sustained resistance. Without external support the People's Opposition would wither in isolation. Without internal receptivity the efforts of the free world would indeed become pointless and unnecessary attempts at "interference."
>
> Given, however, a favorable relationship of internal and external components, and given the right moment when in the course of international events the Soviet Union finds it less painful not to intervene than to intervene, *then* coordinated mass opposition becomes coordinated mass action actively supported by the "Opposition Center" outside.
>
> This "favorable confluence of events" lies in the future—quite possibly in the distant future. Neither the internal nor the external components are yet prepared to take advantage of any unlikely turn of events which might make non-intervention a realistic possibility. But it is clear that every day that passes without some form of increment in the strength of the internal opposition is a day which must be counted in favor of a *Pax Sovietica* in Eastern Europe. That is why the first steps must be taken now to assist the people to act in unison, why an action program, with limited but realistic goals,

must be introduced and firmly planted in the minds of those who form our audiences, and on whom we rely and who rely on us.[15]

Getting this kind of a message across is a difficult propaganda task. It involves subtle and sophisticated reasoning. Subtlety and sophistication are not the strong points of any propaganda. But more than that was involved. The answer to the question, "Why an *action program?*" is based on a specific interpretation of world events in the second half of the twentieth century. The acceptance of RFE's answer was predicated on an acceptance of RFE's view of the world. In short, the fundamental task of VETO was to get the Czechs and Slovaks to accept a certain *Weltanschauung*. Creating a picture of the world of the future and an appreciation of the forces at work to build that world was one of VETO's tasks.

The leaflets and broadcasts designed to carry out this task are an excellent example of how psychological operations attempt to achieve foreign policy goals through influencing the listeners' apparent world. There had been changes in the material world. The West was rearming; it had intervened in the Korean conflict. Riots had occurred in Pilsen and Berlin. People in Czechoslovakia knew about these events. RFE wanted to give them a new meaning and a meaning which would in turn support certain kinds of behavioral reactions among the listening audience.

It was during this second phase that the full power of radio was brought to bear. In addition to carrying the substance of the VETO message, the radio suggested things that could be done to make the People's Opposition a living reality. Broadcasts informed the Czechs and Slovaks, for instance, that the Ballot of the People's Opposition was not to be used in the official election. It could be affixed to walls and bulletin boards, mailed to party members and candidates for election, slipped under thresholds. In short, its ballot box was all of Czechoslovakia.

Programs also reminded the people (and the electoral commissions) of all their legal rights in the elections. The regime had promised that the voting would be free and secret. Election results were to be announced as soon as known, and in all cases where candidates did not receive a majority of the votes

[15] *Ibid.*, p. 2.

new elections were to be held. (It is obvious that RFE did not expect the law to be enforced in all cases. It was making a propaganda demand within the framework of the regime's own legalistic election guarantees designed to make the regime pay dearly in terms of loss of prestige for everything it gained by tampering with the electoral process.)

Although the elections did not give the people a choice between a number of candidates, they had the legal right of crossing out the names of candidates they objected to. Thus RFE advised the voters to examine carefully the lists of candidates. Any candidate that could be of direct or indirect use to the cause of the People's Opposition, should be supported. The hard Communist or crude opportunist should be struck from the list. People should go to the polls with a pencil or other instrument with which to exercise their rights if the opportunity presented itself.[16]

Care was taken to answer the possible response of the defeatist, i.e., that the regime would announce the results of the elections according to its own needs and that any manifestation of opposition was pure futility. This may well be the case, said RFE. But even if the regime does not admit the strength of the opposition, it will know that it exists and how strong it is. Individual candidates will know if they suffer the wrath of the people, and all this will contribute to the wavering and hesitancy the regime has shown since the death of Stalin.

Programs directed at special groups—labor, women, farmers, youth—used the themes that were developed after the June riots (when they were appropriate). Workers and peasants were told of the potentialities of their strength; regime and party officials were encouraged to cross the "golden bridge" or, if this was impossible, to escape; police and the military were appealed to on grounds of their basic loyalties to the Czech and Slovak people. All were reminded of what the Ten Demands meant for them personally.

The first two phases of VETO were laid out before the operation got under way. It is only proper to speak of phases after that in terms of RFE shifting its emphasis and modify-

16 As events unfolded, the electoral issues became much less important than RFE had anticipated because both shop committee and National Assembly elections were postponed until the fall.

ing its tactics to fit the changing conditions and situations in Czechoslovakia. From July through September, for instance, Operation VETO was tied in with the "Harvest of Self-Defense" campaign that RFE waged during the harvest season. The Harvest of Self-Defense dealt with the peasants' struggle against the requisitioning process. In 1954 this was done within the context of the Ten Demands, with particular emphasis on demands Nos. 5 and 6, which concerned agricultural policy, and demand No. 8, which asked for more goods (foodstuffs) for the people.[17] When heavy rains and floods inflicted severe damage in July, RFE reminded its listeners of the way natural catastrophes had been handled by democratic Czechoslovak governments in the past and described how Western governments were dealing with similar situations. It demanded that the regime make compensation for individual losses and reduce the delivery quotas after an honest and just appraisal of the storm and flood damage.[18]

What might be called phase 4 of VETO was in reality a repetition of the tactics that had been planned and only partially used in the spring because of the postponement of the elections. The election campaign in the fall gave RFE a chance to draw on the experience gained in the application of VETO concepts in the spring and to refine techniques and adapt new ones. The policy posture toward the Czechoslovak National Assembly was essentially the same as that developed in the spring: namely, that the parliament is merely a rubber stamp and since the people have no means of influencing the elections or the candidates, it would not be advisable to try to develop an action campaign. The campaign was used as a hook on which to handle programs designed to discredit the parliament further. Scripts were written, for instance, describing the operations of democratic parliaments (Britain, United States, Sweden) and how they functioned as a bulwark against the assumption of dictatorial powers by any one group and how they arrived at social and political programs as the result of free debate and study.[19]

[17] "CS Target No. 17: For Our Campaign re National Harvest of Self-Defense," June 30, 1954.

[18] "CS Target No. 18: Target for Our August Broadcasts," July 28, 1954.

[19] "CS Guidance No. 14: Guidance on the Autumn Elections," September 15, 1954, pp. 1, 4–5.

The Slovak National Council was treated in a different manner. Without any references to Czech-Slovak antagonisms, the Council was attacked as merely the agency which legitimatized decisions made in Prague; in other words, it was merely an instrument of "Communist centralism" that frustrated rather than fulfilled Slovak demands for self-government within a federated Czechoslovak republic.[20]

But it was in regard to the shop committee elections that the *action program* of VETO was truly implemented in the fall of 1954. A number of speeches given in July by Joseph Tesla, chairman of the URO (Revolutionary Trade Union Movement), indicated that perhaps the appeals of RFE to the workers were worrying Communist officials. On July 2, 1954, Tesla gave a speech in which he said:

> All cases of "social democratism" in Union activity lead to protection of unjustified low norms, to various theories of egalitarianism, to the misuse of bonuses. . . . A trade union organization yielding to such opinions will yield to the pressure for wage increases not based on higher productivity. . . . The danger of "social democratism" is always evident where Union officials . . . succumb to the elementary pressure of employees. . . .[21]

Other newspaper stories indicated a similar concern. For example, "In the Metz works in Stepanov and the ZPS works in Lisen, Union officers directly opposed the norm setters in order to protect low norms and unjustified high earnings."[22]

Such official comments reassured the policy advisers at RFE that they were on the right path. They felt that although it was true of Czechoslovakia (as of other countries behind the Iron Curtain) that the peasants were the most dogged opponents of the regime, the allegiance of the industrial worker was nevertheless more important to the regime than subduing the peasants. It was this prize that was being contested in the shop committee elections.

During the pre-election phases (i.e., before the regime announced a definite date for the elections) RFE in the name of

20 *Ibid.*, pp. 1, 5.

21 "Operation VETO," *News from behind the Iron Curtain*, III, No. 10 (October, 1954), 19. The Tesla quote is translated from *Prace*, July 3, 1954.

22 *Ibid.*, quoted from *Prace*, July 6, 1954.

the People's Opposition demanded that the promise made in the spring to hold fall elections be fulfilled. The time had come to begin the return to industrial democracy. Nobody had to teach the prewar Czechoslovak worker anything about democratic trade unionism, but for the benefit of the younger men, RFE broadcast a "course" in trade unionism based on Czechoslovak and Western experience. The scripts were calm in tone, expository and not polemical.[23]

After the elections were announced, the "Workers' Opposition" based its approach on three basic points:

1. That no good can come of the elections if the ROH (Trade Union Federation) puts up candidates known to be its stooges; workers must have freedom to examine the list of candidates in advance, in each mine and factory, and to discuss the list at a plant meeting.
2. That the Shop Council's powers and functions shall be printed and distributed to all workers and that workers shall be free to assemble with candidates and examine them on their knowledge of the powers and functions and their interpretation thereof.
3. That elections shall be truly democratic; candidates to be drawn from among workers in the mine or factory; secret ballot; election by simple majority; mine or factory's own workers as tellers of the vote.[24]

These three points were made a part of the basic doctrine that the function of shop councils is to represent and protect the interests of the workers.

During the whole campaign two basic cautions were observed. No instructions were given on the selections of candidates or on how to vote. There was no talk of "infiltrating" or "capturing" the councils. It was felt that, apart from the fact that it might put individuals in jeopardy of being charged with "acting in obedience to instructions from the Western radio," such comment was not necessary. Individuals in factories knew much more about the specific situation than did RFE and were aware of the kind of action that could be undertaken. RFE looked at its task as perceiving "what the people themselves are doing, [making] the values and meaning of their own acts

23 "CS Guidance No. 14: Guidance and the Autumn Elections," September 15, 1954, p. 8.

24 *Ibid.*, p. 9.

clear to them by looking farther ahead than they are looking and [formulating] their demands [not RFE's] in a concise, easily remembered language which will persuade them that we are truly their brothers because we say exactly what is in their hearts."[25]

The second caution observed was that violent language and strong attack were reserved for those people that the workers genuinely disliked—the higher functionaries of the ROH who were almost never in the factories. Local functionaries sometimes have the sympathy and support of the workers because of the peculiar pressures that the regime can exert upon them or because they have already crossed the "golden bridge," by seeking to promote the worker's interests.[26]

OPERATION FOCUS

The Hungarian Operation FOCUS was the second large-scale radio balloon-leaflet campaign. Its basic elements were identical to those of VETO. It tried to build up the idea of an internal resistance movement (called the "National Opposition Movement"). Twelve demands of the opposition were propagated, and the campaign was launched during an electoral period. However, there were three important differences between the Hungarian and Czechoslovak situations which led to differences in implementation. First, the New Course had gone further in Hungary than it had in Czechoslovakia. More concessions were actually granted and a program of administrative and economic decentralization had begun. Second, the opposition in Hungary was more diffuse and was more significant in the countryside than it was in the industrial and urban centers. Finally, RFE felt that the Hungarian party apparatus showed signs of greater internal strain than did the Czechoslovakian.[27]

By and large, however, VETO and FOCUS are so similar that it is not necessary here to discuss the detailed implementation of the latter.

25 *Ibid.*, p. 3.

26 *Ibid.*

27 "Hungarian Guidance No. 15: Operation FOCUS," September 8, 1954. See also Free Europe Committee, Inc., *President's Report for the Year 1954*, pp. 14–15.

The Evolution of the Propaganda Concepts of VETO and FOCUS

Operations VETO and FOCUS were developed as *action programs* to frustrate the political aims of the New Course while encouraging listeners to apply what pressure they could to force the regimes into granting more and more economic concessions. The action emphasis was terminated following the reversal of the New Course in early 1955 (symbolized by the overthrow of the Nagy regime in Hungary). In a period of "crackdown" RFE felt it would be futile to encourage people to work for concessions. But the other aspect of VETO and FOCUS—the attempt to get the listening audiences to accept a view of the world which included the prospects of liberation—was continued.

Bulganin and Khrushchev's visit to Yugoslavia in May and June of 1955 provided material which caused RFE to refine its concept of liberation through the operation of a "confluence of forces." This visit was undertaken to heal the breach that had developed in 1948. The price the Russians had to pay was one paragraph in the Belgrade Declaration of June 2, 1955, signed by Tito and Bulganin, which read:

> Compliance with the principle of mutual respect for, and non-interference in, internal affairs, for whatever reason, whether of an economic, political or ideological nature, because of questions of internal organization or differences in social system, and of different forms of socialist development, are solely the concerns of the individual countries.[28]

Two months later Tito indicated in no uncertain terms that he interpreted this paragraph to apply not only to Yugoslavia but also to other Communist states in Eastern and Central Europe. On July 27 he unleashed a biting attack against the Hungarian regime. The essence of his charge was that there were still "Stalinists" in Budapest. "They are still putting people in prison . . . who express opinions freely and approve the statements of Khrushchev and other Soviet leaders" (i.e., the statements made at Belgrade). Similar charges were later directed at Prague and Bucharest.[29]

[28] *New York Times*, June 3, 1955, p. 3.

[29] The official position of the United States appears to be contained in a statement Secretary Dulles made to reporters at Brioni on November 6, 1955. The Secretary said that Marshal Tito and he had discussed the "problem of

Radio Free Europe felt that the Belgrade Declaration and Tito's application of its principle to all the captive countries could be a tremendous centrifugal force in the satellite system. Its vast information-gathering agency had provided evidence to show that a number of high Communists in the captive countries would like to have more freedom from Moscow. This conclusion was reinforced by RFE's correspondent at the second Geneva Conference in the fall of 1955, who got the impression that non-Russian Communists were "disgruntled and disillusioned" by the intransigence of the Soviet delegation. They had expected, he wrote, that the Russians "were going to let loose of them a bit and let them manage their own affairs a little more." Molotov, he added, "must have shocked many of those people badly."

Radio Free Europe further reasoned that the Belgrade principle might conceivably influence the relationships between the party leaders and the rest of the populace. Prior to the summer of 1955 every satellite regime was partially protected from local demands (both from within and outside the party) by the general knowledge that the "Kremlin would not allow it." But if Moscow says that it respects the principle of "non-interference in internal affairs" and recognizes "different forms of socialist development," then the regimes stood in a different position both in relation to Moscow and to their own people.

While the Belgrade Declaration provided a foundation on which RFE could carry the propaganda wars to the enemy following the official abandonment of the New Course, other events in 1955 forced RFE to assume a somewhat defensive position. The Summit Conference in Geneva in the summer of 1955 was hailed in some parts of the world as signaling the end of the Cold War. This interpretation supported the Soviet position in the captive nations because it meant that the Western powers had agreed to a permanent Soviet hegemony in Eastern Europe. The photograph of the beaming, genial Eisenhower shaking hands with the respectable, goateed Bulganin could be interpreted by the peoples in the satellite countries as a symbol of an agreement that would seal their fate

---

the States in Eastern Europe and were in common accord in recognizing the importance of the independence of those states, non-interference from outside in their internal affairs, and the right to settle the social and economic order in ways of their own choice." *New York Times,* November 7, 1955, p. 7.

for generations. And the UN membership "package deal" at about the same time, which gave the puppet governments of the satellites a legitimate place in the UN, could be used as further evidence to support the thesis that the captive peoples were a pawn sacrificed to perpetuate the status quo. Fortunately for RFE, President Eisenhower said on August 24, 1955:

> Eagerness to avoid war—if we think no deeper than this single desire—can produce outright or implicit agreement that the injustices and wrongs of the present shall be perpetuated in the future. Thereby we would outrage our own conscience. In the eyes of those who suffer injustice we would become partners with their oppressors. In the judgment of history we would have sold out the freedom of men for a pottage of a false peace. . . . The domination of captive countries cannot longer be justified by a claim that this is needed for purposes of security.[30]

RFE used this statement to support its claims that the U.S. government would agree to no Soviet peace terms that involved an acceptance of the status quo in Central and Eastern Europe.[31]

While the Geneva conferences and the Belgrade Declaration were providing themes for all RFE stations in 1955, a situation known as the "thaw" was developing in Poland that furnished important programing material for the Voice of Free Poland. The debates on culture during this period will be reviewed here briefly, because they offered a convenient opening for fresh attacks and because they demonstrate the difficulties encountered by a totalitarian regime as it tries to control every aspect of human existence. A failure in one area sometimes exposes weaknesses throughout the entire structure.

The debates on culture were touched off by a relatively insignificant event in 1953—the publication by the Institute for Literary Research (IBL) of a work entitled *Literary Periodicals in the Kingdom of Poland in the Years 1832–48*. In a review of this esoteric work in an equally esoteric journal, an author and critic, Chalasinski, questioned one of the fundamental tenets of Marxist-Leninist theory. He questioned the

30 *New York Times,* August 25, 1955, p. 10.

31 RFE Guidance, "Geneva, Belgrade and the Captive Nations," August 15, 1955.

existence of two distinct currents in every national culture. The review did not go unchallenged, and soon Chalasinski found himself accused of liberalism, of attempting to revise "sacred" teaching, and of an eclecticism inspired by bourgeois ideology.

In February, 1955, the Department of Social Sciences of the Polish Academy of Science organized a conference in Zakopane, attended by notable academic and literary personalities. By this time Chalasinski's position was based on three major points, summed up by an RFE researcher as follows:

(1) A disciple of historical materialism is not in duty bound to accept every thesis propounded by the so-called classical marxist thinkers but, basing himself on the theories formulated by Marx, he can develop a notion of the "Existence which determines consciousness" which will answer the requirements of those aspects of cultural life in the examination of which he is engaged. *The validity of supplementary hypotheses is decided on by facts and not by the agreement with this or that statement made by one of the classical marxist thinkers.*

(2) Historical materialism must be supplemented if it is to supply the base for a full and complete theory of culture which will provide a comprehensive explanation of man's activities and of the products of those activities. Historical materialism must also be supplemented if it is to become an adequate "Weltanschauung" explaining to the individual the meaning and purpose of life. Such a "Weltanschauung" must also deal with values which cannot be derived from the teachings of marxism-leninism. Values are drawn from historical experience and from the cultural achievements of mankind from which marxism-leninism has cut itself off and to which *a return must be made if marxism-leninism is to become a comprehensive theory dealing with all the aspects of human thought and action.*

(3) Marxist humanism is in decline. It is characterised by: "Scholarly incompetence, a lack of understanding of relevant issues, or the downright falsification of the true picture of bourgeois learning." [Our italics.][32]

Until June, 1955, the debates on culture were restricted to a very small group of intellectuals. In that month, however, Zolkiewski published an article in *Nowe Drogi* attacking the lit-

[32] Z. Jordan, "Debates on Culture during the 'Thaw'" (RFE-Munich working paper).

erary revisionists and pointing out that their "rightest warpings" not only called for a revised basis for artistic and scientific formulations, but, in the last analysis, amounted to a political attack that struck at the very foundations of the people's democracy and of the dictatorship of the proletariat. This article and others following in *Przeglad Kulturalny* reopened the debate with Chalasinski and his followers, but this time the audience was much larger. And in this debate before a larger audience the defenders of orthodoxy were on the defensive. One of the defenders closed his article with these words: "I regret if I have been unable to convince you. I recognize that I am sometimes short of arguments. This does not mean, however, that I intend to give way."[33]

The Polish desk followed these debates very closely and reported to the Polish people the truly momentous issues that were raised by a critical review of an obscure work on literature. It pointed out to intellectuals the debilitating consequences of a state-controlled culture and indicated how the unanswerable arguments of the intellectual dissidents undermined the rational foundations of the entire Communist society.[34]

If a number of events in 1955 caused RFE to assume a somewhat defensive posture, the Twentieth Congress of the Communist Party of the Soviet Union in February, 1956—and the de-Stalinization campaign—enabled RFE to seize the initiative. Never in its history had RFE been presented with an issue that was so favorable for engaging the enemy on his own ground, for searching out "his vulnerable places, and [strike] them at the foundation of his being." So many themes and variations were used during the ensuing months that it is impossible to repeat them all here. However, they seem to fit into three interrelated general categories: (1) adding fuel to the fire—that is, applying the logic of de-Stalinization in areas where the Communists had never intended them to apply; (2) indicating that the monstrous Stalin was in fact a product of

33 *Ibid.*, p. 8.

34 Although these debates on culture went further and continued longer in Poland than in the other satellites, there was a "revolt" of writers in Hungary in late 1955 and early 1956 which revolved generally around similar issues and which RFE exploited in its broadcasts to that captive country.

the Soviet system and that the only method of insuring that another Stalin will not come to power is to change that system; (3) pointing out that the West was not being deceived by the de-Stalinization program and that it recognized that the basic aims of Soviet imperialism have not indeed been fundamentally altered.

Literally dozens of themes fit into the first category. Radio Free Europe reminded its listeners that editorials and speeches really did not go far in repudiating Stalin. His face still spoiled hundreds of public places, buildings, and books. Everything from mountain peaks to factories had been named after him. To make the record straight all these should be removed and books rewritten. But more important, how about the "little Stalins" who still rule in the satellites? Should not their faces —and with their faces, their influence—be removed from the places of power? Indeed, since Stalin touched every aspect of human existence, should not every aspect of human existence feel the effects of de-Stalinization? A major policy guidance stated the problem this way:

> To stimulate a greater variety and intensity of legitimate and realistic demands, we should ask what stalinism means, what it is that has now been denounced, rejected, and the removal of which our peoples have now been promised. In each country, our peoples have the right to look at what is most abusive and ask, "Is this not stalinism? If yes, when is it to be abolished?" When are we to have opposition parties? A free press? Free elections? Dissolution of collective farms? Free trade unions? More and better consumer goods? Freedom of travel? Freedom of communications with the free world—Western radio, journals of news, opinion, learning, science? Open trade with the West? Is it now to be denounced? . . . Teheran, Yalta, Potsdam: Were they Stalin's personal work? Was the brutal denial of free elections, as demanded by the Potsdam arrangement, Stalin's work? When the United Nations condemned certain "Satellite" regimes for maintaining armies in excess of treaty rights, and for disregard of treaty clauses on human rights, was it condemning stalinism? Will the forces not be decreased and the human rights now be respected by the new anti-stalinists? What is there in the mode of governing our people, the mode of exploiting farmers and workers, the mode of poisoning their children's minds, the mode of preventing them from learning what is going on in the free world and even in the communist world—what is there in all

this that is not stalinist? Tell us, the people, and let us judge for ourselves.[35]

The themes which hammered home the point that since Stalin was no more and no less than a product of the Soviet system, the logic of de-Stalinization demands revising the system in order to make it impossible for any new Stalin to arise, were not mentioned explicitly in the New York guidances but were developed at Munich. The propaganda point to be made is obvious. If Stalin was the monster Khrushchev makes him out to be (and Khrushchev should know), then Khrushchev and all other Communists are obligated to alter the system that created the monster.

Finally, RFE felt that it was important to let the captive peoples know that the West was not being taken in by the apparent changes in the Soviet Union. Diplomatically, this meant that it would not grant concessions to the "new" Soviet Union which would enable the Communists to tighten their strangle hold on the satellites. Domestically, it meant that Socialist parties throughout the world would not yield to Communist demands for "popular front" governments; that the decanonization of Stalin was a Trojan horse which created confusion among the Communists but was met with skepticism by the Socialist parties.[36]

35 "Special Guidance No. 26, Twentieth Congress CPSU," March 27, 1956, pp. 9–10.

36 See "Special Guidance No. 26 (Part 2), Twentieth Congress CPSU," April 7, 1956, pp. 1–13.

• CHAPTER VIII

Conclusions

Case studies provide the kind of richness of detail that is necessary for an understanding of psychological operations but that will always be missing in an analytical treatment of the subject such as was contained in Part I. The three presented were chosen primarily for this purpose, but they can also be used to support our critique of the American approach to psychological operations. This critique has two major points: (1) There is no clearly articulated and widely accepted concept of the nature of the psychological instrument of statecraft. (2) There is no basic and enduring commitment of psychological operations as an integral and necessary part of the foreign policy of any great power in the twentieth century.

Support for this two-pronged criticism may not be apparent to the reader of the case studies who becomes engrossed in the details of operations or who is not aware of the way in which they fit into the pattern of postwar American foreign policy. The American psychological operation in Italy during the Second World War was the first American effort in which the problems of tactical operations against combat troops, appeals to the civilian population behind enemy lines, and the "information" program of a military government in occupied territory were all involved. Operations during the 1948 Italian elections were the most intense undertaken by the United States in the absence of open warfare. By some crude pragmatic measure these operations could be considered successful. The Fascists were ousted from power in Italy; the status of Italy changed from that of enemy to cobelligerent; a group of democratic, free-world–oriented decision-makers were brought into power and American efforts helped keep them in power at a crucial moment in world affairs. Does not the magnitude of the effort indicate commitment? Does not the success indicate an acceptable concept of the nature of the psychological instrument and an acceptable minimum of coordination? Let us probe a little deeper.

The Italian scene in the spring of 1948 was precisely the kind of situation in which the inadequate American conception of psychological operations does not become apparent. The issue in regard to the general election itself was fairly simple and the oversimplified American view of the issue in terms of communism and anticommunism did not deviate disastrously from reality. The straightforward, unambiguous goal of preventing the Communists from winning a general election was clearly understood by all. American aid and the Communist coup in Czechoslovakia in February, 1948, provided a basis for the successful use of tactics of enlightenment and fear with nothing more than a simple, intuitive, common-sense understanding of what was involved. The favorable image of America that was widespread among the Italian people simplified immensely the problem of communicating with and affecting the behavior of the Italian electorate.

To the degree that the success depended on these factors, it was largly fortuitous. Careful study reveals that there was no clear understanding of the psychological operations demonstrated by responsible officials. The thinking of some went little further than a fuzzy notion of a contest between two gladiators with the "minds of men" at stake. The action of others suggests that the contest for Italian voters was seen as something akin to the competition between soap manufacturers for the sale of their products. (Some form letters from Americans, for example, were not far removed from the kind of product indorsements so common in American advertising.) Fortunately (or unfortunately, depending upon one's point of view), there is some similarity between selling soap and winning votes, at least in one specific election campaign. But when the goal is not obvious and short-term, when the issues are not simple, and when the audience is not receptive, much more is needed.

The inadequacy of the operation becomes even clearer if

one examines the way in which the intelligence function was viewed. There is little evidence of a set of organizing principles on the basis of which information was analyzed and presented to the planners. There was only an unstated and vague realization that certain audiences were to be sought for certain messages, but only in rare cases did this lead to the preparation of special messages for special audiences.

It is interesting that a private organization, Radio Free Europe, in its Operation VETO demonstrated a much more refined view of the intelligence function. The policy guidances and the scripts themselves clearly indicate that RFE viewed some audiences as more important than others and realized that special appeals were needed to reach special audiences. For example, the workers were considered to be the most important audience. The Communist officials themselves constituted the second most important audience while peasants and youth groups were given lower priority. The reasoning behind this ordering was not entirely explicit, but it clearly related to the strategy which RFE had devised to achieve its objective. There is some indication that the criteria we have labeled cruciality and susceptibility entered in. It was recognized, for example, that the peasants were the most bitterly antiregime audience, and by implication, the most susceptible. However, because workers were considered more important to the regime, they could have more effect on the activities of official decision-makers. In our terminology they were far more crucial and only slightly less susceptible than the peasants and thus a more important audience. But even this is inadequate. A maximum of rationality can be introduced into the selection of audiences only if the criteria of selection are made explicit and explicitly justified.

Operation VETO also illustrates some other refinements that were absent in the Italian cases. It is clear, for example, that the planners were aware of the "judo" approach to psychological operations. In the programs directed at Communist leaders, little attempt was made to get them to accept a new ideology. It was almost as if Khrushchev's later comment that Communists would give up their beliefs when "shrimps learned to whistle" was accepted by RFE. But this did not remove the Communist leaders from the focus of broadcasts. It was believed possible to affect their behavior without necessarily shaking their beliefs. When RFE's observer at the Geneva Conference in 1955 reported that the leaders of the satellite states wanted freedom from Moscow to pursue more independent paths, RFE tried to strengthen that desire. And the strengthening of that desire had a definite place in RFE's strategy.

The potential for "judo" operations in Italy might well have been great. Perhaps American propaganda could have been used to needle Togliatti or other high-ranking Communists into taking action which would cost them votes. But when operations are viewed in terms of simply winning votes or battling for men's minds, there is no basis for even conceiving of the possibility of this kind of approach.

That there may have been an inadequate conception of psychological operations one might concede, still arguing, however, that the magnitude of the operations reported in the case studies at least demonstrates a commitment to their use. Certainly, there was and is a commitment, but it is of a limited kind. During the war in Italy a sizable and impressive psychological warfare capability was built up. It was achieved primarily because of two factors: (1) General Robert A. McClure's intense belief in "psywar" and the dedication of his small but competent staff and (2) the respect General Eisenhower had for McClure and the support he gave to psychological warfare when most officers were highly skeptical. But the commitment was personal and transitory rather than institutional and permanent. McClure and his small band struggled against strong prejudice or great indifference and won a reluctant permission to operate during the war. When the war was over, the Army's psychological warfare capability deteriorated until today it hardly exists. It has no acceptable place in the military establishment and no strong personal supporters of high rank.

The 1948 operation in Italy indicates a similar kind of commitment. The goal was determinate and immediate and a sizable campaign could be organized. More was involved than just an information program. The four techniques—propaganda, rumor, person-to-person communication, and symbolic acts—were used, although there is little evidence of the deliberate use of rumor. The tactics of enlightenment, terror, and reassurance were also employed. Deception was not used (properly, we believe), and fear would in this case be a better word than terror. The Italian electorate was certainly encouraged by American psychological operations to fear a Communist victory. But this promising beginning did not become the basis for developing a concept of the nature of psychological operations that went beyond "information." The demonstrated effectiveness of properly publicizing American aid did not lead to a systematic exploitation of this technique in the planning and execution of later aid programs in other parts of the world. The commitment to use the psychological instrument lasted until the election was won. In the reorganization of our information services in 1953 (see chap. i) there was no recognition that the United States needed anything but an "information" program. It was as if nobody remembered the 1948 campaign in Italy, or profited by that experience.

It has often been argued that American foreign policy suffers from a lack of goals. We cannot agree that this is its most serious weakness. Goals may not be clearly formulated and articulated, but they do exist. A greater weakness is the absence of total strategies—well-thought-out plans of action to reach goals. It is in the strategies that the roles of the various instruments of statecraft are spelled out. In the absence of total strategies there is no basis for integration. But the development of a role for any instrument of statecraft requires a useful concept of the nature of the instrument. As long as American ideas about the psychological instrument of statecraft are limited to some vague notion of providing information or of cracking the Iron Curtain with a battering ram of truth, total strategies cannot be developed. In this bipolar and power-oriented age we desperately need imaginative and positive action in the direction of the total integration and coordination of *all* our potentials of strength. Without it we may continue to "muddle through"—and we may not. It is hoped that this book has laid the foundations for developing a more adequate and useful concept of strategic psychological operations and the role they can play.

• INDEX

• INDEX

PRINTED IN U.S.A.